A
HISTORY
OF
RADIO PAKISTAN

INSIGNIA, MOTTO AND LOGO

Radio Pakistan's insignia was designed by the well-known artist, Abdur Rahman Chughtai. It is composed of the figure of an eagle in flight, the wings spread out in the form of a crescent enclosing a star, giving it a two-fold association. Firstly, the crescent and star are symbols of Islam, and secondly it incorporates Iqbal's concept of 'Shaheen' as the independent spirited bird with a keen eye, lofty flight and divine attachment. In the original insignia, the words 'Radio Pakistan' were inscribed in Urdu within the crescent. These were replaced by Quranic words in 1951. The *'Ahang'*, Radio Pakistan's programme journal, carried the insignia for the first time with the Quranic inscription and the motto in its issue for the second fortnight of August 1951.

The Arabic script inscribed within the Radio Pakistan insignia are words from the Holy Quran, verse (Ayah) No. 83 of *Surah Al-Baqarah* (speak fair to the people)* and form the motto of Radio Pakistan. These Divine words have been a guideline for Radio Pakistan both in the concept and execution of its policies and programmes. A former Director General of PBC, Qazi Ahmad Saeed, selected these words at the behest of Z.A. Bokhari who wanted a Quranic verse to be adopted as Radio Pakistan's motto. Another claimant to the honour of choosing the motto is M.H. Shah, a former Administrative Officer of PBC. The calligraphy of the motto is the work of Muhammad Yousuf Dehlavi. Syed Saleem A.Gilani, a former Director General of PBC contributed the logo:

NATIONAL ANTHEM

Radio Pakistan has the unique honour of having the National Anthem rehearsed and recorded at its Karachi Station. The artists who performed the one minute and 20 second National Anthem include Najam Ara, Kaukab Jahan, Tanweer Jahan, Shamim Bano, Fazal Elahi, Mashuq Ali Khan, Muhammad Husain Mistri, Buland Iqbal and Daim Husain. Radio Pakistan aired it on August 13, 1954, and played it in the national hook-up every evening to allow people to get familiar with the graceful notes of their National Anthem.

---

* This is a translation by Abdullah Yusuf Ali and Muhammad Farooq-i-Azam Malik. Marmaduke Pickthall's translation of the same text is 'speak kindly to mankind'.

A History of Radio Pakistan

Nihal Ahmad

OXFORD
UNIVERSITY PRESS

Great Clarendon Street, Oxford OX2 6DP

Oxford University Press is a department of the University of Oxford.
It furthers the University's objective of excellence in research, scholarship,
and education by publishing worldwide in

Oxford New York

Auckland Cape Town Dar es Salaam Hong Kong Karachi
Kuala Lumpur Madrid Melbourne Mexico City Nairobi
New Delhi Shanghai Taipei Toronto

with offices in

Argentina Austria Brazil Chile Czech Republic France Greece
Guatemala Hungary Italy Japan South Korea Poland Portugal
Singapore Switzerland Thailand Turkey Ukraine Vietnam

Oxford is a registered trade mark of Oxford University Press
in the UK and in certain other countries

© Oxford University Press 2005

The moral rights of the author have been asserted

First published 2005

All rights reserved. No part of this publication may be reproduced, translated,
stored in a retrieval system, or transmitted, in any form or by any means,
without the prior permission in writing of Oxford University Press.
Enquiries concerning reproduction should be sent to
Oxford University Press at the address below.

This book is sold subject to the condition that it shall not, by way
of trade or otherwise, be lent, re-sold, hired out or otherwise circulated
without the publisher's prior consent in any form of binding or cover
other than that in which it is published and without a similar condition
including this condition being imposed on the subsequent purchaser.

ISBN 0 19 597870 6

Typeset in Times
Printed in Pakistan by
Mehran Printers, Karachi.
Published by
Ameena Saiyid, Oxford University Press
Plot No. 38, Sector 15, Korangi Industrial Area, PO Box 8214
Karachi-74900, Pakistan.

Contents

Preface   vii

1. Introduction — 1
2. Broadcasting in South Asia up to August 1947 — 6
3. This is Radio Pakistan — 12
4. Programme Policy — 25
5. Programme Makers and Programme Trends — 32
6. Change of Status and Nomenclature — 43
7. Programming — 52
8. Overseas Broadcasts — 108
9. News and Current Affairs — 116
10. Planned Development (Engineering) — 129
11. Radio and the Nation in War and Peace — 152
12. Structure and Organization — 180
13. Finance — 204
14. Commercial Service/Sales — 209
15. Central Productions — 213
16. Pakistan Broadcasting Academy — 219
17. Audience Research — 225
18. Press Public Relations, Publications and Overseas Liaison — 230
19. Pakistan Broadcasting Foundation — 238
20. The Future — 241
21. Stations — 252

APPENDICES

I. Press Communiqué of Government of British India — 271
II. Press Note of Government of British India — 273
III. Recipients of PBC Awards, 1998-1999 — 275
IV. Recipients of PBC Awards, 1999-2000 — 279

TABLES

1. Radio Pakistan at a Glance — 281
2. Directors General — 282
3. External Services up to 13 August 1997 — 283
4. External Services effective 14 August 1997 — 285
5. World Service — 286
6. National News Bulletins — 287
7. Summary of News Bulletins — 288
8. Broadcast Centres (Radio Stations) in Pakistan — 288
9. Fixed Transmission Schedule — 290
10. Staff Position — 291
11. B.R. Licences – Issued/Renewed — 292
12. Broadcast Receiver Licence Collections — 293
13. Total Income and Expenditure — 294
14. Material Available in National Sound Archives — 295

Chronology — 297
Bibliography — 305
Index — 307

Preface

**Quaid-i-Azam's Message to the Nation
On the Inauguration of Pakistan Broadcasting Service**

It is with feelings of greatest happiness and emotion that I send you my greetings. August 15 is the birthday of the independent and sovereign State of Pakistan. It marks the fulfilment of the destiny of the Muslim nation which made great sacrifices in the past few years to have its homeland.

At this supreme moment my thoughts are with those valiant fighters in our cause. Pakistan will remain grateful to them and cherish the memory of those who are no more.

The creation of the new State has placed a tremendous responsibility on the citizens of Pakistan. It gives them an opportunity to demonstrate to the world how can a nation, containing many elements, live in peace and amity and work for the betterment of all its citizens, irrespective of caste or creed.

Our object should be peace within and peace without. We want to live peacefully and maintain cordial and friendly relations with our immediate neighbours and with the world at large. We have no aggressive designs against any one. We stand by the United Nations Charter and will gladly make our full contribution to the peace and prosperity of the world.

Muslims of India have shown to the world that they are a united nation, their cause is just and righteous which cannot be denied. Let us, on this day, humbly thank God for His bounty and pray that we might be able to prove that we are worthy of it.

This day marks the end of a poignant phase in our national history and it should also be the beginning of a new and a noble era. Let us impress the minorities by word, deed and thought that as long as they fulfil their duties and obligations as loyal citizens of Pakistan, they have nothing to fear.

To the freedom loving tribes on our borders and the States beyond our borders, we send our greetings and assure that Pakistan will respect their status and will extend to them its most friendly co-operation in preserving peace. We have no ambition beyond the desire to live honourably and let others live honourably.

Today is *Jumat-ul-Wida*, last Friday of the holy month of *Ramazan*, a day of rejoicing for all of us wherever we may be in this vast sub-continent and for the matter of that throughout the world. Let the Muslim congregations in their thousands, in all the mosques, bow in all humility before the Almighty and thank Him for His eternal kindness and generosity, seeking His guidance and assistance in the task of making Pakistan into a great State and themselves into its worthy citizens.

Finally, let me tell you, fellow citizens, Pakistan is a land of great potential resources. But to build it up into a country worthy of the Muslim nation, we shall require every ounce of energy that we possess and I am confident that it will come from all whole-heartedly.' *Pakistan Zindabad*

The above message was issued and broadcast on 15 August 1947.

This book traces the development of radio broadcasting in Pakistan from pre-Partition days to the start of the second millennium. In all probability it is a first major attempt to compile an authentic record of one of the country's pioneering institutions for academic reference and lay interest.

It is not an exhaustive study of broadcasting in Pakistan and there may be details omitted for space considerations. Readers' comments and suggestions are welcome and may be accommodated in future editions.

To add to its reference value, the first chapter in the book gives a brief history of broadcasting from the invention of wireless technology. It offers a comparative study of radio broadcasting in several countries where a different system of management and programming prevailed.

In compiling and writing this book, the main challenge has been availability of authentic material. In this respect, my association with this organization, which spanned four decades, has proven to be an asset. I am grateful to all those who helped me acquire relevant material. I would particularly like to mention Qazi Ahmad Saeed, former Director General of PBC, who made available a couple of old reports and an article written by him on broadcasting. Also my gratitude to Mansur Said, Head of Press and Public Relations Section at PBC Headquarters, for his invaluable cooperation in accessing material from the Corporation's records. Thanks are also due to Mansoor Zafar for his help in the preparation and editing of the manuscript.

1

INTRODUCTION

Radio, or 'wireless telephony' as it was known in the early days, is one of the most fascinating human inventions. It is universally recognised to be among the cheapest and most pervasive media of mass communication. As such, even with the advent of television and the Internet, radio is still considered a powerful means of communication and an easily available source of entertainment, information and education. To date, there are about thirty thousand radio stations in the world, about one-third of which are located in the United States of America.

BEGINNING OF WIRELESS TELEPHONY

The early history of wireless (radio) communication is shrouded in ambiguity and there have been many claimants to its invention. Mahlon Loomis, a dentist of Washington DC, patented a wireless system as early as 1872. In 1864, a Scottish mathematician and physicist, James Clark Maxwell put forward the theory that an electromagnetic signal could be dispatched without any physical link to the point of origin. In 1873, he published a paper entitled 'A Treatise on Electricity and Magnetism' stating that electromagnetic wave energy must exist. Using mathematical equations he demonstrated that electricity, light and heat are essentially the same; they all radiate at the same speed in free space and that this energy is invisible to the human eye.

Maxwell's theories were corroborated by Heinrich Hertz, a German physicist, in a series of experiments conducted between 1887-88. The fundamental unit of frequency, the Hertz, is named after the latter. Actual demonstrations of this wireless theory were conducted by a French national, Edouard Branly in 1891 and a Russian professor, Alexander Popoff on 7 May 1895. The Russians regard Popoff as the inventor of radio and observe 7 May as Radio Day. A British scientist, Sir Oliver Lodge, patented his system in 1897. Nathan B. Stubblefield of Kentucky, USA, demonstrated his wireless telephone system in 1892. However, it is Guglielmo Marconi, an Italian, who is credited with the invention of the radio. He patented his device in London in 1896.

The wireless equipment was first used on ships for navigational purposes. The British Royal Navy and the US Navy first began radio tests in 1889, followed by the German Navy in 1900. The first naval use of radio in a war situation was during the Russo-Japanese war of 1904-5. The superiority of the Marconi equipment used by the Japanese is credited as one of the reasons for their victory over the Russians.

In 1895, Marconi transmitted a wireless telegraph message over a distance of a mile during initial experiments. In 1899 he transmitted across the English Channel. On Christmas Eve of 1906, Reginald Fessenden, a Canadian, transmitted a voice and music programme from a transmitter at Brant Rock, Massachusetts. Charles D. Herrold began broadcasts in 1909 from San Jose, California, and later scheduled regular programmes. Lee de Forest broadcast the Wilson-Hughes presidential election results in 1916. The same year, David Sarnoff, another contender to radio invention fame sent a memorandum to his chief in the American Marconi Company proposing innovative ideas for the development of radio programming. This came to be known as the 'music box memo'.

Radio stations were set up in Montreal (Canada) and the Netherlands in November 1919. Spain and Mexico went on air in 1921, France and the Soviet Union in 1922, Germany in 1923 and Italy in 1924.

The First World War (1914-18) spurred the development and growth of the radiotelephony industry and hastened the introduction of wireless technology in various applications. Broadcasting began within two years of the end of the First World War. By 1926, there were about 170 stations in Europe, 5 in Africa, 40 in Latin America, 10 in Asia, 75 in Canada, 11 in Mexico and 20 in Oceania mainly Australia, New Zealand and the Philippines.

United States of America

In 1920, the first regular broadcasting station in the world was opened by Westinghouse in Pittsburgh (USA). However, several other claims to this honour were made including Montreal (Canada) and the Netherlands.

UNITED KINGDOM

In England, the first radio programmes were broadcast by the Marconi Company from Chelmsford on 23 February 1920. However, it was not till November 1922 that the British Broadcasting Company, with John Reith as its Managing Director, began regular programming.

The British Broadcasting Company was dissolved on 31 December 1926 and the British Broadcasting Corporation (BBC) was constituted under Royal Charter on 1 January 1927. The period from January 1927 to June 1938 has been called 'The Golden Age of Radio' or 'the Reith era'.

The BBC operated under certain restrictions. Advertisements or sponsored programmes could not be aired without Government permission. Subliminal messages were banned. Broadcast of Parliament proceedings were mandatory.

BBC's domestic operations are financed from the sale of licences and profits from BBC enterprises and publications. It also receives payments from its broadcasts of Open University programmes. External Services are financed by annual parliamentary grants-in-aid.

TURKEY

Turkey was one of the first five countries in the world to establish regular radio programmes. The first of these scheduled programmes were broadcast in 1927 from two 5 KW transmitters based in Ankara and Istanbul. The radio stations were initially established and operated by a private group, The Turkish Wireless Telephone Company. This Company managed the two stations till 1937, when they became a state monopoly.

Government administration and monopoly in broadcasting continued till 1961, when the new Constitution stipulated that radio and television broadcasts in Turkey be administered by an independent body. On 1 May 1964 the Turkish Radio and Television Corporation (TRT) was set up.

Along with its General Headquarters in Ankara, TRT has eight regional production centres for radio and television in Ankara, Istanbul, Izmir, and for radio only in Cukurova, Diyarbakir, Erzurum, Trabzon and Antalya.

INTERNATIONAL BROADCASTING

The Netherlands became the first country to broadcast regularly to audiences beyond its own borders, when it inaugurated short wave programming to the Far East in 1927. By the end of 1939, Russia, Italy, Britain, France, Germany and Japan were broadcasting to foreign audiences.

The US Foreign Information Service made its first direct broadcasts to Asia in December 1941 from a studio in San Francisco. On 24 February 1942, Voice of America (VOA) went on air with its first broadcast, a 15-minute programme in German. This was followed up with similar programmes in Italian, French and English. VOA expanded rapidly during the Second World War and by 1945 was broadcasting in forty-one languages. Relay stations were set up around the world to improve the broadcast signal.

During the War, in addition to VOA, the US Government established three radio services in Europe. Radio Free Europe, a CIA project run under the guise of private enterprise, directed its programmes at audiences in Eastern Europe. Radio Liberty, which also claimed private status, was directed at the Soviet Union. RIAS (Radio in American Sector of Berlin) was run by the State Department and was directed at audiences in East Germany. The American Congress legislated the setting up of Radio Marti to broadcast to the people of Cuba in May 1985.

VOA programmes are now broadcast in forty-seven languages including English, for over 157 hours a day or over 1100 hours a week. It claims to have about twelve million regular listeners around the world. VOA programmes broadcast from local radio stations in Asia, Africa, Europe and Latin America are extremely popular.

BBC started external broadcasting on 19 December 1932 with the inauguration of the Empire Service. The aim was to extend domestic broadcasting to listeners overseas, rather than an attempt to unify the diverse peoples of the British Empire.

External Services broadcast programmes in forty-one languages, including English, for over 100 hours a day, using transmitters located within the United Kingdom and on relay bases located from Ascension Island in the Atlantic Ocean to Singapore. Recorded BBC programmes are supplied to many foreign radio stations.

BBC's World Service is the main English programme and runs twenty-four hours daily. In addition to news and current affairs programmes, the Service also features programmes of sports, drama, music and light entertainment. Export promotion is a key function of External Services.

Germany was one of the first countries to begin international broadcasting. Deutsche Welle, like VOA, was the only national radio station financed solely by federal government funds. It has statutory obligations to produce and present programmes for consumption in foreign countries giving 'foreign audiences comprehensive and well-balanced picture of political, cultural and economic life in Germany'. It is also required by law to offer its programmes to everyone interested free of charge.

Deutsche Welle has a fourteen-hour daily service in English, focussing on three target regions—North America, Africa and Asia. For Latin America, it has started offering digital programmes since June 1996. DW airs programmes in German round the clock directed at Europe, North and Central America, Latin America, Australia and New Zealand, Japan, East Asia, South Asia and Africa. It also puts out programmes in thirty-five foreign languages.

2

Broadcasting in South Asia up to August 1947

British India had its first radio broadcast in August 1921 when 'The Times of India' collaborated with the Posts and Telegraphs Department to transmit from its Bombay office. The broadcast was a special music programme for the benefit of the governor of the province, who heard it at Poona, at a distance of about 280 kilometres.

Thereafter the development of broadcasting in India has been somewhat erratic. As in many countries, broadcasting in British India began in the twenties as experimental efforts of private enterprise. The stress was mainly on the sale of receivers rather than on programme content.

The Radio Club of Bombay began transmitting programmes in June, 1923 on a small Marconi transmitter, followed by the Radio Club of Bengal, Calcutta, later that year. The Madras Presidency Radio Club, formed in May 1924 went on air on 31 July 1924, but closed down in 1927 due to financial difficulties. The other two Clubs also eventually suffered financial problems. The Madras Corporation resumed regular service on the same transmitter on 1 April 1930. This Service continued up to 16 June 1938 when All India Radio (AIR) started operating on both medium wave and short wave transmitters.

Indian Broadcasting Company

The 'Indian Broadcasting Company' (IBC) was formed in March 1926 as a limited company. Under an agreement with the Government of India, signed on 13 September 1926, radio stations were to be set up in British India. Government control over the broadcasting service was exercised through the Directorate General of Posts and Telegraphs that functioned under the Department of Industries and Labour of the Government of India. The Bombay Station of IBC went on air on 23 July 1927, while the Calcutta Station was inaugurated a few weeks later on 26 August 1927. Both stations operated on a 1.5 KW MW transmitter with an effective range of 50 kilometres.

The programmes were European and Indian oriented, with a stronger emphasis on European programming and the English language. They included entertainment, mainly music, students' hour, ladies' hour, commercial bulletins and a general news bulletin in English.

INDIAN STATE BROADCASTING SERVICE/ALL INDIA RADIO

The Indian Broadcasting Company soon ran into financial problems and went bankrupt. The reasons cited for its failure were undercapitalization of the company and costly receiving sets. The Company stopped broadcasting in March 1930. The Government of India assumed control from 1 April 1930. 'The Indian State Broadcasting Service' (ISBS), as it was then called, continued with similar programming as the former Indian Broadcasting Company.

The ISBS suffered financial losses during the Depression in the early 1930s, and the shut down of the Service was announced in October 1931. The closure caused public uproar. On 5 May 1932, the ISBS was revived under State management.

The YMCA set up a small 100 watt medium wave transmitter in Lahore in 1928. For two hours every evening it broadcast entertainment programmes and a church service on Sundays. This Service was closed down on 1 September 1937 and a new 5 KW medium wave AIR Station was opened on 16 December 1937.

In January 1935, the Marconi Company loaned the NWFP Government a 250 watt transmitter and community radio sets for use in villages, schools, Government departments and tribal areas. The Provincial Government agreed to purchase the equipment after a year. Aslam Khattak was incharge of programming. The NWFP Government handed over this Station to the Government of India on 1 April 1937.

In 1939, during the Second World War, the Government suspended local programmes and Peshawar became a relay station. Three years later, a broadcasting house with four studios was constructed and a new 10 KW MW transmitter was installed. The Station was formally inaugurated on 16 July 1942.

In 1935, the Indian Agricultural Research Institute in Allahabad began a one-hour daily transmission on a 100 watt transmitter for rural listeners in the area. This ceased after the opening of a 5 KW station at Lucknow in 1938.

Interest in broadcasting was building up in the princely states of Mysore, Hyderabad, Travancore and Baroda. A small station 'Akashvani' was opened by a professor of Mysore University. It was taken over by the Mysore State in 1941 and absorbed in the AIR network on 1 April 1950.

In the state of Travancore, a service on a 5 KW medium wave transmitter was started in 1943 at Trivandrum and became part of AIR in 1950. The Nizam set up two stations at Hyderabad and Aurangabad. Both stations were merged with AIR in 1950. The Baroda Broadcasting Station began in 1939 and merged with AIR in 1948. A small station at Panaji servicing Goa, Daman and Dieu from 1946 was taken over by AIR in 1962.

A separate office of Controller of Broadcasting was created under the Department of Industries and Labour in 1935 and Lionel Fielden of the BBC took over as the first Controller of Broadcasting. With the help of BBC engineers a radio transmission plan was developed comprising of short and medium wave transmitters. The Indian State Broadcasting Service set up stations in Lahore, Delhi, Bombay, Lucknow, Madras and Calcutta from 1936 to 1938.

With the commissioning of short wave transmitters, the first interstation relays between Bombay and Delhi commenced in 1939. Tiruchirapalli and Dhaka Stations were commissioned in the same year. The latter was a complement to the Calcutta Station of AIR and catered to the predominantly Muslim population in the area.

The Indian State Broadcasting Service was renamed 'All India Radio' on 8 June 1936 and in 1943 the title of controller of broadcasting was dropped in favour of director general. A.S. Bokhari became the first Director General of AIR.

With the beginning of the Second World War, transmission hours at all stations were increased from 50 hours 45 minutes a day to 70 hours 15 minutes. Five news bulletins in Tamil, Telugu, Gujarati, Marathi and Pushto were broadcast from Delhi beginning 1 October 1939. These were in addition to the existing English, Hindustani and Bengali news bulletins. The programmes were relayed by other stations through short wave transmitters. At the time of Independence in 1947,

AIR was broadcasting 74 news bulletins—43 in Home Service and 31 in External Services.

India's foreign broadcasts began during the Second World War and became more frequent as the advance of the Axis countries threatened the British Empire. Soon after the declaration of War on 3 September 1939, the Delhi Station broadcasted a Pushto news bulletin aimed at listeners in NWFP and Afghanistan. Soon after, another news bulletin in Afghan-Persian (Dari) began. When Japan joined the War in 1941, high-powered short wave transmitters were ordered to boost transmissions and the first of these five 100 KW SW transmitters was commissioned on 1 May 1944.

By 1945 AIR's external programmes had increased to 74 daily broadcasts in 22 languages. Two years later at the end of the War the number of programmes dropped to 31. By 1990, India's external services broadcast ran for over 90 hours a day in 25 languages.

More changes followed in the early forties. Fielden returned to England when his assignment with AIR ended. Ahmed Shah Bokhari succeeded him as Director General (DG) till the partition of the subcontinent. AIR came under the Ministry of Information and Broadcasting when it was set up in 1941 and by 1943, it was made a permanent organization.

Urdu, Hindi and Hindustani Controversy

While A.S. Bokhari was DG of AIR, allegations were levelled against him of nepotism and preferential treatment of Muslims, and of favouring the Urdu language over Hindi and Hndustani. Even the media picked up on the controversy and nicknamed AIR the 'Bokhari Brothers Corporation' (BBC).

At that time, there was little difference between Hindi and Urdu except that the former was written in the Devanagari script. At one point, Hindu extremist elements insisted that this language be called 'Hindustani'. AIR was perceived to have an anti-Hindi policy and therefore Hindu writers and performers were asked to boycott it.

AIR began using the word 'Hindustani' to avoid linguistic controversy. The Linguistic Survey of India defines this word as under:

> Hindustani primarily is the language of the Upper Gangetic Doab and is also the *lingua franca* of India, capable of being written in both Persian

and Devanagari characters, and without purism, avoiding alike the excessive use of either Persian or Sanskrit words when employed for literature.

The language controversy raged on through the thirties and into the forties becoming particularly virulent towards Urdu programming. Pressure was mounting on AIR to increase Hindi content and Hindu artiste performances.

In an attempt to resolve the language issue, AIR involved distinguished political leaders and scholars in language advisory committees. An important task undertaken by AIR in 1940, on the initiative of Bokhari, was the preparation of a lexicon of 8,000 English words frequently used in its news bulletins, along with the Hindi and Urdu equivalents for each word. The aim was to encourage the use of the most widely understood out of these synonyms. This lexicon was critiqued by scholars of linguistics such as Dr Akhtar Husain, Chiragh Hasan Hasrat, S.H. Vatsayana, Dr Yadu Vanshi and Rafiquddin Ahmed.

The Government of India was forced to get involved in this language controversy and released press communiqués regarding the changing language policy at AIR (for the copies of government press notes see Appendices I and II). AIR used the Hindi language for its programming, particularly its news bulletins. Arabic, Persian and English words were replaced by their Sanskrit equivalents. But the controversy continued as serious objection was taken to the use of this Sanskritised Hindi language by such leaders as Rajagopalachari, Dr Rajendra Prasad, Pandit Jawaharlal Nehru and even Sardar Patel, a protagonist of Hindi and Minister for Information and Broadcasting in the interim Government. They contended that the language had become unintelligible to the majority of the people.

In the heat of controversy the fact that radio as a means of communication should be intelligible to its target audience was forgotten. More than the question of language the real concern was the attempt to deny expression to Muslim musical heritage and talent.

One particularly damaging incident was caused by Sardar Patel, who on assumption of office as Minister for Information and Broadcasting in 1946 fixed the ratio of Urdu, Hindi and Hindustani as 40:40:20, and banned performances by musicians 'whose private lives were a public scandal'. These orders were considered a direct hit at Urdu and Muslim artists.

There was also considerable animosity over the fact that the sensitive post of Director General of All India Radio was entrusted to a Muslim. The Private Secretary to Sardar Patel, Shankar, in his book *My Reminiscences of Sardar Patel* says: 'With Bokhari's departure (on Partition) the greatest stumbling block in the reformation of All India Radio to the new way of thinking was removed.'

3 JUNE 1947 BROADCAST

The year 1947 brought momentous changes with the Partition of the Indian subcontinent. Lord Louis Mountbatten, the last Viceroy of India, announced the Partition Plan on 3 June 1947. On that day, Lord Mountbatten, Pandit Jawaharlal Nehru, Quaid-i-Azam Mohammad Ali Jinnah and Sardar Baldev Singh addressed the Indian people from AIR, New Delhi. This historic broadcast set a seal on the political destiny of 400 million people of British India.

Quaid-i-Azam's cry of '*Pakistan Zindabad*' heralded the beginning of a new state for the 100 million Muslims of the subcontinent. The original broadcasts were made in English, followed by translations in 'Hindustani'.

The partition of the AIR network was inevitable. Of the nine stations, six went to India at Delhi, Calcutta, Bombay, Madras, Lucknow and Tiruchirapalli. The remaining three at Lahore, Peshawar and Dhaka came within Pakistan territory. The entire External Services complex of short wave transmitters, however, remained with India and Pakistan did not receive a share of the equipment. Five other radio broadcasting stations located in the princely states at Mysore, Hyderabad, Aurangabad, Trivandrum and Baroda were also in India and were subsequently taken over by AIR in April 1950. The sixth station at Panaji (Goa) was taken over in 1962. By 1990, AIR had 107 regional stations operating on 143 medium wave channels ranging in power from 1 KW to 1000 KW. In addition, it had 15 FM stations, each on air for 3 to 5 hours. Presently (2000), India has about 150 medium wave, 54 short wave and 110 FM transmitters.

3

THIS IS RADIO PAKISTAN

At midnight of 14 August 1947, Pakistan appeared on the political map of the world as a sovereign state. Radio Pakistan (then known as Pakistan Broadcasting Service) went on the air for the first time. The last announcement of AIR from the Lahore Station was made at 11 p.m. An hour later, the newly composed signature tune of Pakistan Broadcasting Service was played and Zahur Azar made the following announcement:

> At the stroke of twelve midnight the independent sovereign State of Pakistan will come into existence.

At midnight, he made another announcement: 'This is Pakistan Broadcasting Service, Lahore. We now bring to you a special programme on 'The Dawn of Independence'.' It was immediately followed by an Urdu translation by Mustafa Ali Hamadani.

> *Assalamo Alaikum*, Pakistan Broadcasting Service. *Ham Lahore Se Bol Rahe Hain. Raat Ke Bara Baje Hain. Programme Sunye* 'Tulu-i-Subh-i-Azadi'.

The programme included Iqbal's *Milli Tarana* (national song) '*Muslim hain hum watan hai sara jahan humara*' (We are Muslims and the whole world is ours). Similar introductory announcements were made from the Peshawar and Dhaka Stations. At Peshawar, the closing announcement of AIR was made by Yunus Sethi just after 11 p.m. The opening announcement heralding the new Pakistan Broadcasting Service was made immediately after midnight by Aftab Ahmad Bismil in Urdu and Abdullah Jan Maghmoom in Pushto. It was followed by two patriotic songs specially written by Ahmad Nadeem Qasmi and composed and produced by S.S. Niazi. The opening line of the first song was: '*Pakistan banane wale Pakistan mubarak ho*' (Congratulations to the maker of Pakistan).

S.S. Niazi, Ahmad Salman and G.K. Farid were the first Station Directors of Peshawar, Lahore and Dhaka Stations respectively. G.K. Farid took over charge as Station Director Lahore on 1 October 1947 and Capt. Ahsanul Haque became Station Director, Dhaka in his place. Sajjad Sarwar Niazi became the first Station Director of Karachi Station, when it was established on 14 August 1948.

The creation of Pakistan precipitated widespread disturbances and a massive migration of Muslims from minority provinces of India on a scale unprecedented in world history. Pakistan's share of the broadcasting network was very limited. The new country inherited three low-power medium wave regional stations from Undivided India. Of these, two were located in West Pakistan (now Pakistan) at Lahore (5 KW) and Peshawar (10 KW) and one in East Pakistan (now Bangladesh) at Dhaka (5 KW). These three stations formed the new Pakistan Broadcasting Service, and later became Radio Pakistan.

Karachi, the capital of the new country, had no transmitter. There was no short wave transmitter in either wing of the country as none of the short wave transmitters located in Undivided India were given to Pakistan. This made it impossible to centralize the news services and organize national and external broadcasts.

A further aggravation of this lack of resources was that India refused to let Pakistan have its share of frequencies. The international body governing radio frequency refused a new allocation for Pakistan stating it had been issued for the whole of India, including the areas now constituting Pakistan. India was reluctant to share its frequency allotment.

'RADIO PAKISTAN' STATIONS BEFORE INDEPENDENCE

Just prior to Independence, two attempts at initiating a broadcasting service were made in Peshawar and Karachi. As the Congress Government of the NWFP created hurdles for the Muslim League to propagate Quaid-i-Azam's message and canvass for Pakistan for the forthcoming referendum, a young man from Pilibhit, Tahir Husain, the owner of a radio shop in Delhi set up a clandestine radio station for the purpose, with the consent of a Muslim League leader, Sardar Abdur Rab Nishtar.

The short wave transmitter and power supply machine assembled by Tahir Husain was taken to the residence of Sardar Abdur Rab Nishtar in three fruit baskets.

On the evening of 24 April 1946 the first broadcast was made from a house in Peshawar with the opening announcement: 'We are speaking from Radio Pakistan on the 70 metre band'. Begum Mumtaz Jamal was the first reporter and announcer. The new 'station' moved its location frequently to avoid detection. The Provincial Government announced a prize of Rs.10,000/- for the arrest of those involved in the illegal activity. The 'Station' was closed after the Muslim League won the NWFP referendum.

In a similar incident in Karachi, a 'station' was set up to broadcast Independence Day ceremonies. This endeavour, however, had the blessing of the Provincial Government, with T.N. Idnani, Adviser to the Government of Sindh, acting as coordinator. A transmitter was assembled from spare parts purchased from junk shops and fitted in the barracks of Ack Ack School. A studio was also set up in the barracks, where heavy curtains were used to get the right acoustics.

The first experimental broadcast was made on 5 August 1947. This 'Sindh Government Broadcasting Station' began regular broadcasts five days later. It gave live coverage of the historic ceremonies heralding the birth of Pakistan and Quaid-i-Azam's oath-taking as Governor General. The station aired a variety of music and drama programmes. However, ten days later the station was shut down citing a violation of the Wireless Telegraphy Act that states that a Provincial Government cannot operate a radio station. Those who worked for this Station included B.H. Zaidi, S.K. Hyder, S.R. Bhavani, B.H. Syed, Rashid Ahmed Bhatti, Feroz Sulemani, Husain Bakhsh and Qutub. T.N. Idnani led the team.

DEVELOPMENT IN EARLY YEARS

Soon after Pakistan gained independence, a 'priority programme' for radio broadcasting development was undertaken. According to the programme, a medium wave transmitter was to be procured for the Federal Capital, that was Karachi in those days. Short wave transmitters would be set up in Karachi and Dhaka to provide a link between the two Wings of the country, separated by over a thousand miles of Indian territory. Short wave transmitters were needed to initiate an External Services broadcast to act as the 'Voice of Pakistan' promoting the ideology and policies of this fledgling state.

Before this programme could materialize, India invaded the Muslim-majority state of Kashmir. There was a heightened need for information services to boost the morale of the Kashmiri freedom fighters, or Mujahideen, and present their viewpoint to counter Indian propaganda. Pakistan lacked equipment and a short wave transmitter. To make a start, a 500-watt short wave transmitter, used in the Second World War, was obtained from a junk dealer and repaired and put into service. A public address system, microphones and other equipment were purchased and installed on a truck. Two generators were fitted on to another truck.

From this make-shift facility, test transmission began on 10 April 1948 and regular programmes began a week later. In the early days the two trucks were moved around to avoid detection by Indian forces. Then a site about 40 kilometres from the border was selected to serve as a broadcasting station. The two trucks were hidden behind tall pine trees. A room was built to serve as a studio, with blankets hung on the walls to achieve the right acoustics. This was the humble beginning of a station that later came to be known as the Rawalpindi-III Station.

The new station had a clear reception in Peshawar, Lahore and Srinagar. The staff at the station received fan mail from Iran, Saudi Arabia and Kenya. Some listeners in Kenya sent table tennis equipment for the staff working under trying conditions in snow-capped mountains. The staff consisted of four engineers, two news editors and two producers.

In 1949, the service became more professional as a powerful transmitter, studio equipment and two wire recorders were obtained. The 'Station' broadcasts were so effective it became a target of the Indian forces and was bombarded on July 1948, with no damage being caused.

In October 1947, Riaz Ahmad, Chief Engineer, who was on deputation from the Directorate General of Civil Aviation, was sent to the United Kingdom and the United States of America to purchase transmitters and other equipment that was not available locally. Before the Second World War, Holland and Germany were the major manufacturers of transmitters and broadcasting equipments. After the War the manufacturing facilities had either been destroyed or shut down and supply to world markets had ceased. The UK and USA had become the new major suppliers to the world.

Two short wave transmitters of 50 KW each and one 10 KW medium wave transmitter were purchased from America and shipped to Pakistan, destined for Landhi in Karachi. The Chief Engineer went to the USA on a second shopping mission in August 1951 to purchase transmitters and studio equipment for stations in East Pakistan. He made a subsequent trip to Australia for the same purpose.

A small transmitter was set up to serve the Capital City, and was installed in an abandoned dilapidated barrack at the Intelligence School in Karachi. A KMC gutter pipe was used as a mast. In addition to this 100-watt medium wave transmitter, another 250-watt short wave transmitter was set up in the same room. It was hoped this combination would allow coverage of a wider area of Karachi. The studios were also located in the same barrack and the 'station' office was set up in tents. Such was the beginning of the broadcasting organization in the capital of what was then the sixth largest state in the world.

The 'emergency station' was ready in time to commemorate the country's first anniversary of independence. This improvised Karachi Station of Radio Pakistan went on air on 14 August 1948. It was inaugurated by Quaid-i-Millat Liaquat Ali Khan, after *tilawat* and *dua* (recitation from the Holy Quran and prayer) performed by Maulana Shabbir Ahmad Usmani. The Minister for Information and Broadcasting, Khwaja Shahabuddin, presented a talk on the occasion.

An interesting incident occurred during the broadcast. While the Quaid-i-Millat was addressing the nation there was a power failure. The generator went on in the control room and Z.A. Bokhari focussed a torch on the script so that the programme continued without interruption. At the end, the Prime Minister asked why the studio did not use a petromax. Bokhari replied that the studio needed permission from the Ministry of Finance to purchase one and the Prime Minister gave special permission for two petromaxes to be purchased. The Quaid-i-Millat referred to this experience in a speech delivered at the opening of the new Broadcasting House in Karachi on 16 July 1951:

> I remember 14 August 1948 when I went to the temporary studios at No. 6 Intelligence School for my first broadcast from there. There I saw the sand covered faces of people working among those tents and huts. Addressing them, I said then: 'You are beginning from scratch very much like we had to begin Pakistan from scratch'.

He was very hopeful and optimistic about the future of broadcasting in Pakistan when he said:

> The sand and the tents at No. 6 Intelligence School were in some way reminiscent of the early history of Muslims in Arabia and I wondered whether the resemblance would not prove deeper still. The Muslims of Arabia carried the message of Islam to the whole world. Radio Pakistan may also one day make the voice of Pakistan heard all over the world.

A 10-KW medium wave transmitter was inaugurated by the Minister for Information and Broadcasting on 1 November 1948 at Landhi, about 24 kilometres from Karachi. It was set up in a dilapidated building. A diesel electric power plant was fitted since no land lines were available at the time. The staff and artistes had to wade through water during the rainy season to reach the studios. Transmission radiated on 363.6 metres with a frequency of 825 kilocycles. The studios remained in the barracks at the Intelligence School, and the link with the transmitter at Landhi was made through a frequency modulated transmitter.

The Headquarters of Radio Pakistan as well as the Central News Organization had to be stationed at Lahore for a few months before relocating to Karachi. The News Organization was accommodated at Broadcasting House and the Headquarters was set up in a local hotel.

Radio Pakistan suffered severe staff shortages in these early days. The service was left with only ninteen engineers of different cadres to operate the three Stations. Most of the non-Muslim engineers and employees had migrated to India. The rest of the staff who stayed on or who migrated from India suffered great hardship and inconvenience in getting to work because of widespread civil unrest. Office supplies could not be obtained as commercial centres were closed. Staff travelling to Lahore from Delhi was attacked at Amritsar and several employees were forced to reside on the office premises because of the unsafe travelling conditions.

Plans to have news bulletins on 14 August never materialized. There were no arrangements for centralized broadcast of news and a shortage of radio journalists and ancillary staff made it difficult to generate news material. There was a complete breakdown of news agency services and no way to receive or transmit news to the four important cities (Karachi, Peshawar, Lahore and Dhaka) of Pakistan.

The 'Station' managed to get equipment and furniture together to sustain its operations. The local authorities arranged armed escort for the news staff in the early morning curfew hours. It proved difficult to draft the first news bulletin. No accurate reports were received from Karachi on 14 August 1947, the day Pakistan came into being.

Still, Radio Pakistan's first news bulletin went on air on the morning of 15 August 1947 with the words: 'This is the Pakistan Broadcasting Service. Here is the news....' It was followed by an Urdu bulletin. These first bulletins were well received all over West Pakistan, in particular the Urdu bulletin. People were generally pleased to hear their own language, Urdu, after having listened to the Sanskritised Hindustani bulletins of AIR in the last few months before Partition.

The Dhaka and Peshawar News Units also experienced teething problems, the most serious being the absence of creed machines. For several months, Peshawar relied on Lahore bulletins, telephone and airmail messages, and Provincial news reports from local departments. The Dhaka Unit received a creed machine before 14 August 1947 but it remained inoperative as the Indian Government refused to link the Calcutta circuit with Dhaka. On the night of 14 August some editors spent long hours at the Telegraph Office trying to persuade the Calcutta Exchange to switch on the creed circuit. The Lahore-New Delhi creed circuit was also disconnected in the disturbances that followed Partition.

After Independence, the centralized News Organization of Delhi was split into three units located at Lahore, Peshawar and Dhaka, as there was no centralized system available in Pakistan.

Those were very trying days in many ways. In the mayhem that followed the Partition of the subcontinent, there was a great need for this vital information source to maintain an objective and calm approach in presenting the news. It also aimed to change the general mood of fear and despair to one of hope and courage.

SOS Messages

The jubilation of Independence was sadly tempered by widespread rioting, mass killings, arson and looting, rape and abduction of women and attacks on public transport. The genocide spurred a large-scale migration of Muslims from the Muslim-minority provinces of India to Pakistan and similarly Hindus migrated to India. Hundreds of thousands of displaced Muslims sought refuge in their promised homeland. The new Government of Pakistan did not have the resources to meet this emergency situation. The refugees were placed in camps like the Walton Camp that had very little in the way of shelter.

A total breakdown in communication links between Pakistan and India added to the hardship of the people, particularly those lost in transition.

In these circumstances Radio Pakistan became a vital link for lost families and friends and displaced persons trying to contact others for assistance. From 27 August 1947 Lahore Station started a regular service for refugees, broadcasting messages from people trying to locate lost ones. Information on the status of displaced persons was also given. Special staff was hired to receive messages at the gate, on the telephone and at the refugee camps. These were scrutinized and then broadcast. The addresses of persons in India were never given to avoid identification and harassment by Indian authorities. The SOS service continued for about seven months till the end of March 1948. During this period about 36,900 SOS messages were broadcast. It was a humanitarian service that brought relief to thousands.

During the period of the SOS service, the Duty Room of Lahore Station witnessed many scenes that reflected the human tragedy unfolding in the subcontinent at the time. In one incident, a middle-aged man brought a small child to the Duty Room saying that he found the girl sleeping under a bush near the Jallo border. The staff broadcast an announcement of the girl's description every half hour. Shortly afterwards, a distraught man entered the Duty Room and the little girl ran to him calling out *'Abba! Abba!* (father! father!). The man embraced his daughter and then collapsed in exhaustion.

NATION-BUILDING PROGRAMMES

In order to allow for the broadcast of lost persons messages, entertainment programmes had to be discontinued. Music programmes were reintroduced in October 1947, but in keeping with the troubled times, the tone of these programmes was sober.

Several morale boosting programmes such as 'Pakistan Hamara Hai' (Pakistan is ours) were introduced at this time. This series covered topics in line with the Government's nation-building programme. It soon became a popular series respected for its candour and objective views.

In November 1947 torrential rains added to the misery of the people stranded in the Camp, particularly children, the elderly and the sick. The programme 'Pakistan Hamara Hai' drew comparisons between the historic account of aid rendered by the Ansars of Madina to the Muhajirs of Makkah at the time of *Hijrat* and the people of Lahore who showered the refugees of Walton Camp with humanitarian aid in the form of food, medicines, clothing and other essentials. So overwhelming was the support that the Commissioner of Lahore called the Station to thank the donors and requested them to cease donations as the immediate needs of the refugees had been met.

Programmes similar to 'Pakistan Hamara Hai' were started from other stations as well. 'Pakistan Zindabad' (later changed to 'Waqt Ki Awaz') and 'Pakistan Zamung De' were broadcast from Peshawar; and 'Amader Pakistan' and 'Pak Watan Hamara' from Dhaka. When Karachi Station went on the air in August 1948, it featured a programme 'Istiqlal-i-Pakistan' along with its Sindhi version 'Pakistan Asanjo Ahe'. These programmes became very popular and were credited with helping to boost public morale and hope in the future. In September-October 1947 a special series of talks entitled 'Islam Ka Nizam-i-Hayat' by Maulana Abul Ala Maududi was introduced. As the political situation improved, variety programmes featuring discussions and drama were introduced. Actuality programmes and outside broadcasts were also started.

India attacked Kashmir on October 1947 and the hostilities caused an influx of refugees from the Indian-occupied areas of the state into Pakistan. Radio Pakistan began a special daily programme in the Kashmiri language for the benefit of the refugees. This programme continued up to 14 August 1949 when it was merged with a Kashmiri programme originating from Karachi.

DEATH OF QUAID-I-AZAM

On 14 August 1948 Pakistanis celebrated the anniversary of Independence Day. Conditions in the country had begun to stabilize. It was a shock to the nation, therefore, when on 11 September 1948 the Father of the Nation, Quaid-i-Azam Mohammad Ali Jinnah, breathed his last, barely a year after realizing the dream of a separate homeland for Muslims. Losing its beloved leader was a blow to the fledgling state.

Radio Pakistan interrupted its transmission to announce the Quaid's death:

"ہم قوم کو ایک نہایت اندوہناک خبر سناتے ہیں قائداعظم محمد علی جناح کا وصال ہو گیا"

(We regret to announce a tragic news. Quaid-i-Azam Mohammad Ali Jinnah has expired).

The announcement was followed by Quran Khwani and listeners were asked to remain tuned to Radio Pakistan for an update of the news. On 12 September a three-hour commentary on the funeral was broadcast from Karachi Station and carried by Lahore, Peshawar and Dhaka Stations. The funeral procession from Quaid-i-Azam's residence to his last resting place was covered live by commentators stationed along the route. The commentators were Agha Muhammad Ashraf, Z.A. Bokhari and Siddique Ahmad Siddiqi, a well-known Aligarian orator.

This outdoor broadcast conducted under difficult conditions, technical and otherwise, is one of the most outstanding achievements in the history of Radio Pakistan. The same evening Quaid-i-Millat Liaquat Ali Khan addressed the nation from the Karachi Station. A forty-day mourning period began and in keeping with the sombre spirit of a devastated nation, Radio Pakistan altered its programming to include religious programmes.

Assassination of Gandhiji

On 30 January 1948, Mohandas Karamchand Gandhi (popularly known among Hindus as Mahatama Gandhi or Bapu) was assassinated. In its night news bulletin, Radio Pakistan reported on the assassination and presented a life sketch of Gandhiji. A special obituary programme entitled 'Ghoroob-i-Azmat' (End of Greatness) was broadcast on the same day.

Short Wave Transmitters

Radio Pakistan's expansion plans included setting up medium wave and short wave transmitters to link the two wings of the country with a central broadcasting system. On 17 January 1949, a 7.5 kilowatt short wave transmitter was commissioned at Dhaka. This was the first short wave transmitter for Pakistan and it covered East Pakistan and surrounding areas.

At Landhi, Karachi, new installations included two 50 kilowatt short wave transmitters and one 10 KW MW transmitter with emergency studio facilities; a power generating station consisting of two diesel electric power plants of 250 KW each and two 50 KW power. The first high power short wave transmitter of Radio Pakistan was on the air by 14 August 1949, Pakistan's third Independence Day. The new facility was inaugurated by the Governor General, Khwaja Nazimuddin. With the installation of this transmitter, four external services were introduced in Arabic, Iranian, Afghan-Persian and Burmese. A low power (300 watt) short wave transmitter was commissioned on 1 October 1949 at Lahore. The other 50 KW SW transmitter at Karachi was brought into service on 25 December 1949.

This marked the completion of the first phase of the priority programme of technical development for initiating a radio link between the two Wings of the country. Radio Pakistan could now reach out to the Indo-Pakistan subcontinent and serve as Pakistan's window to the outside world. Further development plans were contingent on the availability of frequencies as allocated by the International High Frequency Broadcasting Conference and confirmed by the World Radio Conference during 1950-51.

By 1949, there was a definite improvement in the mood and confidence of the Pakistani nation. Radio Pakistan reflected this optimism in a major revision of its programme policies. Programmes exhorting national reconstruction, projecting Islamic thought and culture and nation-building were introduced to meet the changing needs of listeners. Programmes targeted at foreign audiences were also introduced. Music entertainment made a welcome come-back after the long period of solemnity during the emergency.

The International Islamic Economic Conference held in Karachi from 25 November to 10 December 1949 was given extensive coverage by Radio Pakistan. Interviews with delegates and speeches by experts from amongst the delegates were broadcast every evening in a special IIEC programme.

Radio Pakistan held its third Directors Conference at Karachi in December 1949. Suggestions for programming improvements included emphasis on folk music and poetic talent. The Conference reviewed religious programmes, stressing that they be used for moral uplift and character-building. Tilawat broadcasts were to be centralized. A Station Engineers Conference was also held in 1949. It was the first of its kind to be held in the Indo-Pakistan subcontinent.

The fourth Station Directors Conference was held in Karachi on 7-13 September 1950. Plans were discussed to make Radio Pakistan appeal to the general public, with cultural and religious programmes and news bulletins conducive to easy listening. Measures for detection of unlicensed radio sets and punishment of offenders was discussed.

Throughout 1950 Radio Pakistan continued expanding its service within the country as well as to neighbouring Islamic countries.

Frequency Allotment

For the first time since the Second World War, an International High Frequency Broadcasting Conference was held in Mexico in 1948-49 to reallocate the short wave band to accommodate new countries that had emerged on the map of the world. Pakistan joined the Conference on 22 October 1948 for the first time. Radio Pakistan was represented by A.S. Bokhari, Riaz Ahmad, Director of Engineering and S.A. Aziz, Research Engineer. A model plan was drafted for organized short wave broadcasting of various countries of the world. Pakistan was tentatively allotted 141 channel hours.

Pakistan played a leading role in the Conference proceedings and was elected to several committees including the experts committee—the Technical Plan Committee. This Committee met in Paris in 1949 to work on the model plan and drafted six plans to suit the propagation conditions of the various sunspot 'seasons'. This meeting was attended by Rashid Ahmad, Deputy Controller of Broadcasting and S.A. Aziz, Research Engineer. Radio Pakistan also participated in the Region III Conference and the Provisional Frequency Board at Geneva, which dealt with the medium wave band and the 90 metre band. As a result of the deliberations of this Conference, Pakistan was allotted a workable share of the spectrum in these bands.

The history of Radio Pakistan from the early days of Partition shows the steady development of the organization from a humble makeshift beginning. The pioneering spirit, hard work and perseverance of its management took it from a small broadcasting station with a transmission of 27 hours per day on three low-power medium wave transmitters to its present day (mid-2001) technologically sophisticated organization broadcasting 365 hours programmes from 24 home stations and nearly 22 hours programmes in the World and External Services, a total of 387 hours per day.

The contribution of these early pioneers of broadcasting in Pakistan in establishing the framework of the organization and setting its professional standards has earned them a place in the history books of this country. They rightly deserve to be remembered as the benefactors and architects who laid sound foundations of the organization on which the entire edifice has been built.

Besides the late Syed Zulfiquar Ali Bokhari, who was the main source of inspiration, other key contributors were Syed Rashid Ahmad, Ahmad Salman (formerly Jugal Kishore Mehra), Sajjad Sarwar Niazi, G.K. Farid, Mahmood Nizami, Asnain Qutb, S.A.Hafeez Hoshiarpuri, N.M. Rashed and Agha Bashir Ahmad in programme development; Riaz Ahmad, Bashir Ahmad, S.A. Aziz and M.A. Rehman in the Engineering wing; Mohamad Sarfraz, Sheikh Ihsanul Haque, A. Ghani Eirabie, M.M. Sayeed, K.S.H. Ansari, Hamid Jalal, N.H. Hashmey and Saeedul Haq in the News department.

In recognition of his services to the development of broadcasting in Pakistan, the Auditorium in the Broadcasting House, Islamabad, has been named after Syed Z.A. Bokhari.

4

Programme Policy

Pre-Partition Policy

In the pre-Partition period, broadcasting was under the private sector, but the government had the authority to censor programme content and if necessary shut down a station. Stations were required to broadcast free of charge weather forecasts, Government communiqués and educational material, provided these did not exceed 10 per cent of regular programming time. Speeches and discussions had to be strictly non-political in nature and had to be pre-approved by a government authority. News items had to be composed from press releases supplied by government-approved news agencies.

Government servants could participate in radio programmes with permission. There were restrictions on advertising and advertisements and announcements were limited to 10 per cent of any programme timing.

Content restrictions prohibited offensive material, seditious intent or that which was likely to incite racial or communal animosity. Objectivity in reporting was to be maintained. Government activities and initiatives were to be publicized.

During the Second World War, A.S. Bokhari, then officiating Controller of Broadcasting, stated that broadcasting was a peace-time activity and was popular for its entertainment value and rational reporting. He was against violent and monotonous propaganda being introduced in regular programme content and stressed the need for more entertainment and cultural content to attract listeners' interest. He felt that post-War broadcasting should revert to projecting development activities, raising public taste and fostering artistic talent.

When the Interim Government was set up in 1946, the need to revise these broadcasting rules was realized. In October of that year, revised broadcasting instructions were sent to stations. In brief, they stated:

1. The ban on discussions of party politics or party propaganda and subjects of religious controversy would continue.
2. Broadcasts by Ministers in Provinces or Members of the Interim Government on action taken by them within the sphere of their responsibility not to be regarded as party propaganda, but references whether appreciative, critical or neutral, to individual political parties or to party politics to be avoided.
3. India's right to independence having been recognised constitutional questions about India-Britain relations were no longer a subject of controversy. Questions on constitutional controversy between political parties in India, to be avoided in all broadcasts.
4. Broadcasts by Members of Interim Government, Provincial Ministers and other public persons, whatever the subject of broadcast need not be referred by Stations to the Director General except in doubtful cases.
5. The practice of Viceroy's approval in advance in the case of Governors to be maintained, except in grave emergency. Similarly, broadcasts by the rulers of states need Viceroy's permission.
6. Previous permission of the Ministry of Information and Broadcasting was necessary in the case of broadcasts by Ministers of Indian States. Broadcasts by officials of States did not need prior permission of the Director General, except in doubtful cases.
7. Appeals to the public by employers or employees over trade or industrial disputes not to be broadcast. In the case of strike by Government employees, the Government had the right to explain the necessity or justification of the attitude or action taken by the Government in suppressing disorder or preventing breaches of peace resulting from industrial strikes or threat of such strikes. In these cases prior permission of the Director General was not necessary, except in doubtful cases.
8. The general ban on appeals to the public for raising funds to be maintained.

Programme Policy after Independence

Pakistan's information policy is based on the tenets of Article 19 of its Constitution, which relates to the fundamental right of freedom of speech and expression. It is also based on Articles 31 to 40 of Chapter 2 of Part II, which relate to such principles of policy as the affirmation of the Islamic way of life; promotion of local government institutions; discouragement of parochial and other social evils; promotion of women's participation and social justice; protection of family values and minority rights; and strengthening bonds with the Muslim world. Article 19 states:

> Every citizen shall have the right to freedom of speech and expression, and there shall be freedom of the press, subject to any reasonable restrictions imposed by law in the interest of the glory of Islam or the integrity, security or defence of Pakistan or any part thereof, friendly relations with foreign states, public order, decency or morality, or in relation to contempt of Court, commission of or incitement to an offence.

These Constitutional provisions applied to all media of information, radio, television, press and films. Based on these guidelines, the Federal Government announced an information policy, which included a section on broadcasting. The Minister for Information and Broadcasting addressed the Third Conference of Station Directors of Radio Pakistan in December 1949 and announced the following principles of programming:

1. Radio Pakistan should primarily be an instrument of nation-building, without ignoring healthy entertainment;
2. It should reflect the soul of Pakistan and represent the artistic traditions of its people, their ideals and their aspirations;
3. It should be a powerful unifying factor between various parts of Pakistan, in particular between East Pakistan and West Pakistan, which were separated by one thousand miles;
4. The quality of programmes should not be sacrificed when aiming at any reorientation of programme policy; and
5. It should aim at a close collaboration with various departments of the Government in the Centre and in the Provinces.

The government of General Ayub Khan announced its information policy on 23 October 1965. The policy re-affirmed the principal ideology of the country and committed to adopting modern technology. Broadcasting should serve the national interests. Programming should appeal to general interest rather than that of select groups. The scope of Radio Pakistan's activities should widen to include all sections of society, trades and institutions. The broadcasting policy was aimed at fostering national unity, integration and solidarity. It's aim was to keep alive the spirit of nationalism awakened since the September war with India.

Successive federal governments introduced similar broadcasting policies. In July 1977, a new broadcasting policy stressed freedom of the information media and encouraged objective reporting, even of political and national issues. It exhorted the promotion of Islam as a progressive religion. The media was asked to act as a bridge between the public and government agencies.

Radio Pakistan adopted the following general guidelines to meet the spirit and intentions of government broadcasting policies.

GENERAL GUIDELINES

1. To provide information, education and entertainment through programmes which maintain a proper balance in their subject-matter and a high general standard of quality and morality.
2. To broadcast such programmes as may promote Islamic ideology, national unity and principles of democracy, freedom, equality, tolerance and social justice as enunciated by Islam.
3. To broadcast programmes which discourage parochial, racial, tribal, sectarian, linguistic and provincial prejudices and reflect the national urges and aspirations of the people of Pakistan.
4. To create and enlarge public conscientiousness of the nation's determination to maintain its independence, preserve its integrity, promote its ideology and raise its economic and social level.
5. To inculcate a sense of patriotism among the people of Pakistan so that they take pride in their ideology, cultural heritage and products and place national interests before personal interests.
6. To present Islam as a dynamic, progressive and comprehensive code of life fully capable of meeting the challenges of modern space age and providing guidance and solution to its complex

problems in all spheres. Religious programmes to aim at creating a better understanding of Islam and harmony among different sects, taking care that broadcasts do not injure religious susceptibilities or accentuate sectarian differences.
7. To broadcast programmes projecting development activities and to provide a communication support to on-going development projects in order to ensure the participation of the community in development effort.
8. To project the policies and progress of Pakistan and the various aspects of life of its people with a view to promoting goodwill for Pakistan and building up its prestige among foreign listeners.
9. To provide healthy entertainment and to develop, broaden and raise the public taste, keeping a balance between what the public wants and what it needs.
10. The approach in programmes directed to the educated youth is modern and rationalistic rather than didactic and traditional.

GUIDELINES FOR REGIONAL CULTURES AND LANGUAGES

1. To present the rich variety of regional languages and cultures as part of the national culture so that they supplement each other instead of supplanting; and to broadcast suitable programmes from regional stations but prominence has to be given to programmes in the national language.
2. To use simple, lucid and commonly understood and spoken language and pronunciation in broadcasts. To encourage the use of commonly understood Urdu words in the regional languages instead of looking for archaic words or coining new ones. Similarly, to reflect local intonation and style, make use of words of regional languages in Urdu broadcasts from regional stations.

GUIDELINES FOR MUSIC PROGRAMMES

1. The Muslims of South Asia have made outstanding contributions to the development of classical music which occupies a place of distinction in our national cultural heritage. To preserve and promote this heritage along with other forms of popular and folk music.

2. To avoid scrupulously erotic and vulgar songs whether film or non-film.
3. To evolve through concerted efforts a distinct style of popular music in conformity with our national traditions of dignity and sobriety. To discourage cheap imitation of Western music, while appreciating the need for introducing harmony in our music through orchestration.

Guidelines for Drama

1. Dramatic programmes should be purposeful. To encourage playwrights to write plays on significant events in Islamic history, on the lives of Muslim heroes and Islamic values like truthfulness, self-sacrifice, love for fellow-beings, patriotism, courage, etc.
2. To avoid plays having vulgarity of expression, overt romanticism, uninhibited courtship, too much crime and violence, sob stuff and words or gestures depicting sensuality.

Guidelines for News and Current Affairs

1. PBC to build up itself as a reliable and authentic source of news; and to establish its credibility and confidence among the listeners. To take note that a news service is in the nature of a public trust and the listeners have a right to hear all the significant news, good or bad.
2. To go for immediacy, accuracy and impartiality governed by national interest as the main elements in its news policy.
3. To arrange collection of news of important events through its own correspondents and monitoring of foreign broadcasting stations while continuing to depend on APP, PPI, affiliated foreign news agencies and the Government press releases and handouts for the supply of news.
4. To maintain impartiality in the news coverage of political party leaders; major political parties receiving more coverage than the minor ones.
5. To cover activities of political parties and statements of party leaders, but to avoid personal attacks on one another.

6. Preceding general elections to arrange a series of political broadcasts by leaders of significant political parties explaining their programmes and policies without casting aspersions on other parties. To fix in consultation with the Federal Government the number of candidates a party must enter in order to qualify for political broadcast. Statements made by Ministers pertaining to their official work, the participation of Ministers, MNAs or politicians in discussion programmes and reference to politicians in news bulletins need not be regarded as party political broadcasting.

General PBC Code

The following restrictions must be observed:

1. Undue criticism of friendly countries. However, reference to and/or a passionate and balanced discussion of policies pursued by any of them is not debarred.
2. Anything obscene or defamatory.
3. Incitement to violence or anything against maintenance of law and order.
4. Anything amounting to contempt of Court.
5. Attack on a political party by name.
6. Anything showing disrespect to the Constitution or advocating change in the Constitution through violence; but advocating changes in a constitutional way should not be debarred.
7. Anything against the Law of the land.

5

Programme Makers and Programme Trends

Role of Radio/Instrument of Change

The aim of broadcasting is three-fold, to entertain, to inform and to educate. Some may add a fourth priority, and that is, to motivate and influence public opinion. The priority of each aim differs from country to country. The aims are not mutually exclusive but are interdependent.

There are two widely and divergent approaches. One approach favours the entertainment value of broadcasting over social and moral values. Lionel Fielden, First Controller of Broadcasting, AIR, felt that almost seven-eighth of transmission time should be devoted to music or other light entertainment. This ratio was difficult to maintain even during his time in broadcasting. Spoken-word and serious topic programmes took up over one-third of on-air time. According to Fielden, the end result of catering to the lowest common factor of public taste would be to create 'boring background noise'. He was in favour of raising public taste and appreciation of finer programming without losing the interest of the average listener.

Lord Reith (Managing Director, BBC) was the main exponent of this school of thought. He believed that radio should give listeners more than they anticipated. It should be uplifting. Radio was regarded as an instrument of social change, a means to raise cultural awareness. It is believed to create a climate for change by influencing values and attitudes, and it ushers in modernization. Radio is a vital source of knowledge. It helps change the power structure in traditional or orthodox societies by educating and empowering the masses through knowledge of subjects such as agriculture, health and hygiene. As a popular mass media, its programmes and advertisements reach a greater audience and raises awareness and aspirations of people. Radio contributes to the planning and implementation of development programmes by building awareness of the scope and benefits of these projects.

Radio has also played a major role in developing public appreciation for various types of music from folk to classical. Programme developers have come to realise that they must cater to different tastes and produce high quality programs if they are to sustain and increase their audiences.

QUALIFICATIONS OF PROGRAMME PRODUCERS

A programme producer selects and develops programmes, sets broadcast schedules, and establishes quality standards. He must take into consideration the policies of the organization, the creative freedom allowed to him, the Station's resources in hardware and software. He must know his audience, their tastes and preferences, and the issues sensitive to them. The merits of a broadcasting station rest largely on the creativity and competence of its programming staff. Particularly crucial are the capabilities and qualifications of the programme producer.

The ideal programme producer or developer is imaginative, creative, has good general knowledge, is a good editor, and is adept at working with people. Fielden had once said:

> If India wants good broadcasting she will have to make up her mind that what is needed is not a balance of quarrelsome elements but men with sufficient integrity to rise above class or caste or communal jealousies and direct a cultural medium with intelligence and zeal.

Fielden's choice of programmers is indicative of his views on recruitment. His programme recruits came from varied backgrounds. They were lovers of music, poetry, science and technology, playwrights, writers, actors, and fresh university graduates. There was even a former sugar technologist on staff.

Successful programming personnel are the ones who have the knack of getting the best out of other contributors—writers, orators, intellectuals, poets, singers and actors of different age groups and social backgrounds. They must have exceptional creative and organizational skills to visualize and implement successful programme content.

The planning, production and presentation of radio programmes is a specialised art and requires a rare combination of aptitude and talent. It requires persons with specialized skills.

The ultimate aim in hiring programme producer was to ensure they had the ability to develop popular programmes and increase listenership. The programme producer had to be a 'generalist' rather than a 'specialist' with the ability to handle a variety of programme assignments.

Programme Trends

Change in broadcasting techniques and programme content is essential to keep up with changing interests and needs of an increasingly sophisticated audience. As Syed Saleem A. Gilani, a former Director General, PBC once said:

> The quality and quantity of information, and variety and canvas of current affairs topics in the fast shrinking world of today, as indeed the special requirements, specific to our new and developing nation, put additional demands on the broadcasters to constantly adjust their programme-fare and its presentation.

From the beginning, radio was expected to play a role in promoting development projects, moral and social values, and other issues of national significance. This inevitably increased the ratio of spoken-word programmes at the cost of entertainment, and the need was felt to balance the proportion of entertainment programming with that of information and inspirational programmes.

Under its Charter, PBC is required to maintain a proper balance in programme content. A balance of opinion including controversial or differing viewpoints was to be encouraged, because it fostered listeners' interest.

The Mumtaz Hasan Broadcasting Committee stated that a prime function of radio in Pakistan should be to promote education. Governments cannot tackle the education of its population alone without the participation of the people themselves. Broadcasting has the potential to play a major role in developing a new consciousness of the need for education. The Committee strongly recommended that the importance of entertainment in broadcasting be recognized and given a primary consideration in regular programming. The Committee deplored the cultural vacuum existing in the country at the time, referring to the ban on dance and music in educational institutions. The Committee recommended that a proper ratio of entertainment and public affairs programmes be maintained. They noted an increase in the number of spoken-word programmes over the years and felt that unless a proper balance was maintained, the aims of good broadcasting would be defeated. It considered Radio Pakistan's programme content at that time as not conducive to sustained listening.

The process of change in social conditions and communication methods started with the end of colonial rule in Asian countries. The colonial rulers made few changes in the social norms and values, especially in rural society. In the rural areas, communication was carried out through interpersonal contact or by word of mouth. The village elder, the religious scholar, the school teacher and the hakeem (traditional practitioner of medicine) were the opinion leaders. Their word carried authority among the villagers. Print was an elitist medium and confined to the educated urban class, reaching a mere 5 per cent of the population. The traditional daily gathering at sundown in the village 'chaupal' (meeting place/community centre) was the centre of communication activity. The weekly village market, the seasonal fair to honour a saint and the harvest festivals were the major communication events.

After gaining independence from colonialists, Asian or Third World countries began setting up industrial infrastructure in urban centres, which spurred the migration of rural youth to urban areas. These young people became a source of information and ence posed a challenge to the authority of the village elders. The situation was further aggravated when the rural labour migrated to oil exporting countries to work and earn high wages. They were exposed to the latest communication technologies and media. The traditional rural information channels began to vanish, and there was nothing to fill the information void except the radio.

Radio became the preferred communication medium in a large number of Third World countries because of the inaccessibility of electricity and television. Radio is and will remain the most effective medium for reaching and educating the masses in countries with vast areas, scattered population and low per capita income.

Though radio does not compete successfully with television as a broadcasting medium, yet it has its own compelling merits that ensure a dedicated audience. In countries of the West where television is the main source of entertainment, radio audiences score high in popularity ratings. It is the medium of choice for information and entertainment while travelling, working at home or in the office, for visually impaired persons and for all those situations where watching television is not feasible. Radios are portable, more accessible and cheaper than televisions.

Despite its merits, radio broadcasting in Pakistan has not been given due priority in national planning. Direct satellite broadcasting and television has overtaken radio as the preferred medium of information and entertainment, except in emergency crisis situations. In many countries, as in Pakistan, government financial allocations for radio broadcasting development has not kept pace with the rise in operational costs. This has had an adverse impact on the development of this communication medium.

The decline of radio in a country with high poverty and illiteracy levels raises serious communication issues. Radio has been an affordable and cost effective medium that reaches out to even the poor and isolated segments of society. Radio speaks to target audiences in their own language and in this fact lies much of its communication power.

The fact that radio has played an active role in changing attitudes and educating people through information and entertainment is often ignored by policy makers who are not investing sufficient resources in the development of this medium.

Rapid advances in communication techniques and electronics engineering accentuated the need for a change in broadcast strategies to meet increased competition from other media. Today, radio has to compete for audiences against film, television, video and audio recorders, foreign radio channels and other local stations. Many of the major international networks have considerable technical sophistication and financial resources.

The invention of the transistor in the fifties had a tremendous impact on the size of radio audiences. This cheap and portable electronic device was convenient and affordable. It changed radio audiences from the family or group entertainment environment to an individual enjoyment option as well. The advent of car radios also brought new and larger audiences.

Another significant innovation was the introduction of tapes and tape recording gear in the fifties. Radio stations were able to take the microphone out of the studios into the streets and wherever news-making events were taking place. This eased the problems of outside broadcasting, actuality reporting and voice inserts thereby adding to the authenticity and credibility of programmes as well as of the radio station. This in turn created a sense of participation and involvement for listeners, particularly those in villages and remote areas. More people were turning to radio than any other medium—print or electronic, in the under-developed conditions prevailing in Pakistan, particularly in rural areas.

With the change and unprecedented increase in listenership, a need for change in programming became essential. Changes were made in 1947 and again in 1965. Independence called for a complete reorientation of broadcast strategies to meet the needs of a new nation traumatized by partition violence and lacking basic resources and civic infrastructure. Radio Pakistan had played a conscious role in nation-building.

Pakistan's broadcasters were also conscious of the fact that radio was an instrument of political power, and had great potential for both good and evil. In the past, especially during the Second World War, it had been used as a propaganda machine. Since its foundation, Radio Pakistan has striven to inform and educate and to present Pakistan's viewpoint to the world.

Unlike pre-Partition days when radio was his master's voice, it had now to be the voice of the nation. As Syed Saleem A. Gilani said:

> The founding fathers of broadcasting in this part of the world sat down to the task of creating a radio of an independent nation. A whole new set of programmes was conceived, new priorities were determined, new ratios among information, education and entertainment were fixed and the secular radio of British days with its new found independence adjusted its religious programmes to cater to the needs of the new Islamic nation. Radio talks started highlighting the achievements of Muslims in the cultural field, administrative system of Muslim Governments, codes of Islamic life and morality, works of Muslim historians, scientists, philosophers, reformers and men of letters. There was an interesting interplay between the literature of national language and the vast treasure of literature available in the provincial languages. This had the potential of overall enrichment, cultural proximity, integration and feeling of oneness.

Programming for children was specially formulated to educate and reform. Patriotism, honesty, pride in Islamic history and way of life were inherent themes of children's hour programmes. Music was promoted and a new style of ghazal rendition was developed. Folk music was promoted. Works of old Muslim maestros, vocalists and composers were revived and included in routine music programmes.

New programmes were planned and produced, many of them targeted at the vast Muslim population of the country. Islamic history and culture were presented as a model of learning, together with a discussion on modern day problems facing the new nation.

Radio programming stayed in tune with national events in the early years following Independence, particularly during catastrophes like the influx of refugees, the demise of Mr Jinnah and Liaquat Ali Khan, floods, earthquakes, wars of 1965 and 1971 and the Ojhri Camp disaster. Special programmes were broadcast to provide information updates and emotional support.

Beginning 1972, several changes and additions were made in programmes. Morning transmission was added for farmers from Hyderabad, Lahore and Multan. District and local news had live recordings, with stringers being appointed at some district headquarters. Labour documentary programmes with interviews of factory workers were organized to educate workers on their rights and duties. Haj service programmes were broadcast four months a year. Special World Service programmes in Urdu and English were broadcast for Pakistanis living abroad.

In 1977 the nation faced a crisis of faith in its leadership and national institutions. Radio Pakistan became involved in the task of pulling the country out of chaos. Special programming ran for 200 hours a day in the Home Service in 16 languages/dialects and over 30 hours a day in the World and External Services in 16 languages. Great stress was laid on the process of Islamization. Islam was presented as a dynamic and forward-looking religion and listeners were exhorted to practise the Islamic way of life on a day to day basis. Programmes actively promoting communication between the government and the people were introduced, providing a forum to the people to ventilate their grievances and hear the viewpoint of the Government. These programmes on the one hand involved the common man in the affairs of the country and on the other the Government became more alive to its own responsibilities.

New Concept of Broadcasting

A drastic change in the programme pattern was made in 1987. It was first introduced in the News Wing on 1 November 1986 on an experimental basis. News broadcasts were made every hour on the hour, or every half hour. Current affairs were brought into sharper focus. A change in the style of presentation and news content was also made. The formal style gave way to an informal address. Radio presenters adopted a friendly and intimate style of rapport with listeners allowing for audience participation.

Programme content also underwent a radical change in format. Transmission time was segmented into hourly chunks with each hour managed by a producer and compere. News bulletins in regional languages were featured at every half hour. Short talk shows on topical issues, catering to diverse interests brought in new audiences.

Radio Pakistan became actively involved in community development and uplift programmes. A new concept of service evolved for radio programmes. Talk shows with guidance counsellors helped many individuals with personal problems. Even community problems were taken on. For example, in the Yazman area of the Cholistan desert, large flocks of sick goats and sheep were saved through the timely intervention of Radio Bahawalpur. In Chak 44/NDB in Bahawalpur, the wheat crop was saved.

Through the efforts of Radio Lahore Station, Chak 163 NB of Sargodha received electricity, a girls primary school and boys middle school were upgraded, a veterinary dispensary was set up and one and a half kilometre long metalled link road was built, and canal water was brought to 34 chaks. The people of Chak 163 NB were so grateful for these project undertakings they put up a plaque in honour of Radio Pakistan at the main road crossing. The plaque was unveiled by the then Director General, PBC on October, 1988.

Radio Pakistan became involved in many other community welfare projects, such as blood banks, book banks, drug banks, and even an eye bank. Free medical camps were organized, scholarships for orphans and poor children, wheel chairs were arranged for invalids and handicapped persons. The radio stations got their listeners to participate in these fund raising activities. Free medical camps were organized which catered to hundreds of poor, especially women and children from remote areas who had no access to a doctor, treatment facilities or medicines.

Other instances of the universal influence of Radio broadcasts in the lives of people was an incident in which a *seth* (rich man) of Hyderabad was released by his kidnappers after a news item was broadcast that the son of the seth had died of shock at the news of his father's abduction. In Karachi, a traffic cell was set up at Radio Pakistan in collaboration with local police authorities to control the city's traffic at this busy junction. City Service cells and outdoor broadcasting units were located at all stations for timely radio reports of current events in the city.

An emergency message service has been used to benefit many, in particular notices of missing children under the age of twelve years has been a regular feature on radio. Several incidents of recoveries of missing children are credited to radio announcements. Since 1988, a regular programme 'Yeh Bacha Kis Ka Hai' (Whose child is this?) is featured live on the national transmission every night in cooperation with the Edhi Welfare Centre, Karachi. A child is interviewed daily in an attempt to locate his/her parents or guardians.

The new concept of broadcasting was well-received by listeners, but requests for the return of popular programmes such as dramas, quiz programmes, *musha'iras*, concerts, *qirat* and *naat* competitions, had to be entertained. Eventually programme content evolved into a fine mix of information, education and entertainment.

By 1987, the first anniversary of the new concept broadcasting, thousands of letters and telephone calls were received at all stations. Radio audiences had expanded at a phenomenal rate. The country's postal authorities issued a stamp to commemorate the successful completion of one year of the new concept of broadcasting.

The new programme strategy and format projected development programmes launched by the Government. However, with the change in Government this strategy was modified and became less effective. It was generally assumed that this change was made because the new strategy was in sync with the information policy of the previous Muslim League Government. This point was confirmed by a pro-PPP journalist, Mahmood Sham in the daily *Jang* of 26 March 1989 wherein he pointed out that the broadcast style was '*awami*' (folksy) which suited the aims of the Pakistan Peoples Party (PPP) to have direct contact with the people.

Under the PPP Government, this programme strategy would have been highly effective in conveying the Government's viewpoint, its plans and policies. In Sham's opinion, the change in the broadcast strategy was either a proof of the short-sightedness of the decision-makers in the Ministry of Information and Broadcasting or a conspiracy to misinform the people. He felt that Radio Pakistan had reverted to the old style of broadcasting and lost its direction and goals. In January 1991, Radio Pakistan once again reviewed its programme pattern and strategy and implemented many changes which exist today.

STAFF ARTISTS

Staff artists are a category of employees quite distinct from the regular employees of a broadcasting organization. These positions cannot be filled with regular Government servants as specialized skills and talent is an essential prerequisite. These artists are engaged on short-term contracts as instrumentalists, newscasters, announcers, comperes, copyists etc. They are hired on a monthly basis and are generally known as 'monthly paid artists'. Later these artists began protesting for long term contracts and benefit packages. These were gradually granted. The tenure was extended to one year, and in 1965, the contract period was raised to five years. A number of other concessions were granted, including gratuity, vacation pay, maternity leave, sick leave, travelling and daily allowances and Government residential accommodation.

When Radio Pakistan became a statutory Corporation there were 467 staff artists employed on a contract basis. These contract employees demanded regular employment through a trade union and their demand was accepted. A disadvantage of this decision was that staff artists, who are generally employable as long as their performance and popularity with listeners endures, were given long-term service contracts similar to government servants. Hence they could not be replaced with fresh talent when their popularity ratings dropped.

Another decision taken at the time of regularising staff artists was to freeze hiring in that post so that when it fell vacant it was not filled. So in effect these positions became a dying cadre. As a result, at the end of June 2000 there were 143 persons on the pay roll against the sanctioned number of 463. The broad categories of staff artists are:

1. Music Accompanyists
2. Music Vocalists
3. Music Composers
4. Music Staff Notation Supervisors
5. Western Music Supervisors
6. Drama Producers
7. Drama Voices
8. Drama and Feature Scriptwriters
9. Sound Effects Men
10. Compilers of Songs
11. Copyists and Calligraphists
12. Urdu Typists
13. Pronunciation Experts
14. Publicity and Reference Assistants
15. Supervisors Gramophone Records Library
16. News Translators
17. Newsreaders
18. Qaris
19. Announcers
20. Special Audience Programme Comperes

6

CHANGE OF STATUS AND NOMENCLATURE

NEED FOR CHANGE

Radio Pakistan made good progress in the first two decades of its existence, but still lagged behind international developments in broadcasting. There were two basic reasons for this. Firstly, in the national development planning, broadcasting was not given due priority. Secondly, there was poor administration in the organization. It was realised in the early sixties that the lack of administrative, financial and editorial freedom was a hindrance in the efficient management of Radio as a function-oriented public service medium.

The Government has attempted on several occasions to have a thorough review of the functions of Radio Pakistan to suit the philosophy of Pakistan and conduct a reorganization of its structure to meet its growing needs. Committees and commissions were appointed to conduct the reviews and reorganization. For example on 1 November 1959 the Government set up a Radio Pakistan Planning Board that thoroughly examined and evaluated all programmes broadcast from Radio Pakistan Stations. The Board produced a report suggesting changes in programming and expansion of engineering facilities. In the light of these recommendations, the full range of programmes broadcast from Radio Pakistan in both the Wings of the country were reviewed and extensive changes were implemented.

MUMTAZ HASAN COMMITTEE

The Government appointed the most important of these broadcasting committees on 29 July 1965. This eight-member Committee under the chairmanship of Mumtaz Hasan was to examine and report on the state of broadcasting, and make recommendations and guidelines for its development as an instrument of mass communication and mass education. Other members of the Committee were: Akhtar Hameed, Begum Mariam Hashimuddin, Prof. Munir Choudhry, A.K. Sumar, Syed Imtiaz Ali Taj and Dr A. Waheed. The Director General Radio Pakistan was an ex-officio member.

The following were the terms of reference of the Committee:

1. To examine and report on the state of broadcasting in Pakistan and to make recommendations regarding the lines of its future growth and development and in particular:
    i) to consider the role of broadcasting in a developing country like Pakistan, its goals and objectives in the light of national ideology and aspirations, and its responsibilities as a medium of education, information and entertainment;
    ii) to examine the type and quality of programmes offered to the general listeners in the Home and External Services of Radio Pakistan and to recommend measures for enhancing their effectiveness in relation to the objectives in view;
    iii) to review the existing programmes for special audiences, e.g., the programmes addressed to students, agriculturists, women and children, industrial labour and other key social groups and to suggest such changes as may be necessary;
    iv) to examine the existing arrangements for Listener Research in Radio Pakistan and to recommend ways and means of placing it on a more scientific basis;
    v) to survey and report on the present listening facilities in the country and to examine measures for extending listening facilities;
    vi) to review the present system of licensing of radio sets and the collection of revenue and to suggest anti-piracy measures;
    vii) to examine the organizational structure of Radio Pakistan and to suggest any changes which may be conducive to better technical operation and more professional programme production;
    viii) to examine the adequacy of programme, engineering and news personnel and to consider measures necessary for attracting to and retaining in the service of Radio Pakistan persons of the required calibre and qualifications;

ix) to examine the terms and conditions of employment of 'Staff Artists' and to suggest such improvements in these terms and conditions as may be necessary;
x) to review the working and operation of the Commercial Service of Radio Pakistan and to recommend lines of the future policy and scope of commercial broadcasting;
xi) to examine the working of the Central News Organization and the Regional News Units along with the present arrangements for the collection, compilation and editing of news, and to suggest measures for streamlining these arrangements; and
xii) to make a survey of technical facilities at the Broadcasting Houses and other installations and to examine the conditions of transmitting and other equipment, and suggest measures for their adequate and economic use as well as for expansion and improvement of facilities, where necessary.
2. The Committee would suggest guidelines for the development of Television as an instrument of mass communication and mass education.
3. The Committee could also make such other recommendations as may be necessary for the development of broadcasting on sound national lines consistent with the country's ideology, socio-economic development and financial resources.

The Committee functioned as an independent body, but required expert consultation on broadcasting from Radio Pakistan. A special cell was, therefore, set up at the Headquarters to collect professional literature on broadcasting in various countries, to gather information on payroll and administration and programme content from various foreign radio organizations. The committee would also seek public opinion and input.

The formal secretariat of the Committee was set up in February 1966, and began extensive research work on radio and television broadcasting within the country and in several countries of the West. The Report of the Broadcasting Committee was submitted to the Government on 28 April 1967. Its major structural recommendation was to set up an autonomous broadcasting corporation that would be responsible for both sound broadcasting and television in the country. The proposed corporation was to be established by an Act of Legislature that should specify the degree, extent and nature of Government control over its policy and objectives. The Corporation should be an autonomous public corporation run by a five-member Board of Governors and would have an independent administration, but still be under the supervisory control of the Government, particularly in matters of policy. The Committee recommended that the Government should transfer all assets of Radio Pakistan to the proposed Corporation, and continue with the annual allocations to the Stations.

The Committee felt that Government control of broadcasting should gradually decrease. Similarly it recommended that the Government should transfer its interest in the Television Promoters Company to the proposed Corporation, after negotiating terms with the two foreign participants, the Nippon Electric Company and Thomson Television International. By the middle of 1968 when either of the contracts would be renewable by the participants, the question of foreign participation should be reconsidered.

The Broadcasting Committee recommended that the Corporation be headed by a President with three deputies for sound broadcasting, television, and administration and coordination. These officers should be professionals from within the field. Pay scales at Radio Pakistan were to be upgraded and made equivalent to junior Class One (i.e. grade 17), and streamlined for both Television and Radio. The disparity in salaries for similar positions was to be removed. Newsreaders and External Services staff were to be engaged on special terms. Improved pay scales for all categories of staff artists and fees for casual artists were also suggested by the Committee.

In its report the committee put forward a number of arguments in support of its recommendations, which are summarized as under:

1. The present set-up whatever its utility may have been before independence cannot, in our opinion, provide he kind of service which is expected of it in a free and developing country.
2. A broadcasting organization, if it has to fulfil effectively the basic objectives of providing education, information and entertainment, cannot afford to be run as just another department of the Government. The controls and procedural formalities which may perhaps be necessary in the effective administration of routine Government activity, when applied to a broadcasting organization, fail to achieve results. Radio Pakistan which, in our view, can be the most effective agent in helping to bring about an educational and cultural renaissance in the country is hampered in the

performance of this role by bureaucratic controls. This is particularly evident in the matter of financial sanctions and recruitment of personnel. The rules which control Radio Pakistan's personnel and their activities are no different from those operative in the case of purely administrative or revenue departments. The requirements of a growing and sensitive organization like Radio Pakistan cannot readily be appreciated or expeditiously dealt with in the normal routine.
3. The present salaries and opportunities offered to Radio personnel are at par with those of other departments belonging to centrally constituted services. The result is that Radio Pakistan does not get the type of talent it needs for maintaining proper standards of creative work. The point of entry into the organization has to be made respectable enough to attract the right kind of persons. At present the pay and status of the officers of Radio Pakistan are not good enough to give them that certain degree of security and confidence which is so necessary for effective functioning.
4. Radio Pakistan is subject to stringent financial rules which tend to curb its initiative as a professional organization.
5. The argument that in a developing country like Pakistan it would be unwise for the Government to surrender management of a vital medium like the Radio to a body largely outside its control, does not appear to be valid. Even if Radio is not a Government department, but a statutory non-government organization, it would still have to function under specific controls and be responsible to the Government and the people in every respect. Everywhere in the world the State exercises a broad and general supervisory control on broadcasting. The charter which confers corporate status on a broadcasting organization defines clearly the nature of Government control and the limits of the Corporation's autonomy.
6. It is true to some extent that a non-government organization can be open to political and commercial pressures but proper safeguards can always be provided in this respect in the organization's charter. Wherever broadcasting organizations are run by statutory bodies, they have been found to be sufficiently objective, fair and beyond the reach of unhealthy pressures.
7. The argument that even as a non-government organization Radio Pakistan will have to be heavily subsidised by the Government and therefore it may as well stay with the Government, is fallacious. Good and responsible broadcasting is public service

and since every activity of a Government is directed towards this end, the status of the body which the Government subsidizes hardly matters as long as it competently performs the functions for which it has been established.
8. Radio Pakistan has limited operational flexibility. Therefore, it has not been able to build up the level of prestige that was due. Its image needs to be improved.
9. As early as 1936, it was decided by the old British Government in India that the future of broadcasting lay in a public corporation.
10. In the case of broadcasting what Pakistan needs is an autonomous public corporation which should be internally and administratively and financially independent, but, by and large, under the supervisory control of the Government, particularly in matters of policy.

ADMINISTRATIVE COMMITTEE

The Government appointed an Administrative Committee consisting of the representatives of the Ministries of Finance and Information and Broadcasting, to examine the Broadcasting Committee Report. The Administrative Committee completed its job on 29 November 1967. It supported most of the recommendations including the suggestion to establish an autonomous radio corporation. The Administrative Committee's comments in this connection are reproduced below:

> If Radio Pakistan is to fulfil effectively the basic objectives of providing education, information and entertainment, then it is necessary that it should be free from rigid controls and procedural formalities which perhaps are necessary in the administration of a routine Government activity, but inhibit the performance of a sensitive medium like Radio. The activities of Radio Pakistan are predominantly of a creative nature and the present administrative and financial controls are serious obstacles in its way to achieve the desired goals. Radio Pakistan has embarked upon ambitious programmes in various fields. It is projecting effectively the policy of the Government and is striving very hard to promote national integration, solidarity and unity. It is endeavouring through a variety of its programmes to create eagerness and willingness among the listeners to actively participate in the huge task of national economic uplift. It is felt that unless the quality of staff employed by this important organization is of the

requisite standard and calibre, broadcasting may not be in a position to have the desired impact on the people. The present scales of pay do not attract talented people and there are cases of many outstanding young men of having left the organization to take up more lucrative posts elsewhere. At present because of lack of administrative and financial powers it is difficult for Radio Pakistan to show the best results. The recommendation made by the Broadcasting Committee for the establishment of a Corporation is, therefore, timely and is strongly recommended.

The other reason which now makes it necessary to support this recommendation is the establishment of the Pakistan Television Corporation. Television is a competitive medium and as a result of its becoming a Corporation it has many specific advantages over Radio Pakistan. In the first place, the power to hire and fire, which it can exercise and which Radio Pakistan lacks, would enable it to have better and more talented staff. Secondly, better grades of pay and other facilities have enabled Television Corporation to motivate its employees. This had adversely affected the morale of the staff of Radio Pakistan and already many members of the staff have made efforts to leave Radio Pakistan and join Television Corporation. The third factor which has given Television an advantage over Radio Pakistan is the better scales of fees that are offered to the artistes. Hitherto, Radio Pakistan had considerable hold over the artistes, but more attractive fees have enabled Television to attract artistes. The same artistes are no longer as eager to perform for Radio Pakistan as before. Although it is a desirable development from the point of view of artistes, yet this disparity is a handicap for Radio Pakistan and is bound to affect the quality of its programmes.

Another disadvantage for Radio was that it was rapidly losing its urban audiences to television. This further aggravated the situation for Radio staff who had difficulty in competing with the glamour of Television. They needed the incentives of better pay scales, better service and working conditions and greater freedom of action, as was the case with Television employees. It was felt this could be achieved by making Radio Pakistan a Corporation.

In other countries, wherever broadcasting is run by an autonomous authority, television and radio have flourished. This has resulted in valuable adances and development in programme techniques and technical knowledge. These organizations have been successful in winning large audiences both at home and abroad and have served their nations well. Broadcasting organizations in the United Kingdom, Australia, Canada, Japan and other countries are examples of successes.

Radio Pakistan would similarly do well if its current organization was changed and it was made an autonomous body.

Before Radio Pakistan could be incorporated, the Ministry of Finance hired a consultant to prepare a feasibility report of the financial implications of such a move. The report of the consultants on 'Possibility of Converting the Government Department of Radio Pakistan Into a Public Corporation' proved the economic viability of the project and was approved by the Ministry of Finance.

The actual implementation of the recommendation to make Radio Pakistan a corporation ran into delays. Finally, more than five and a half years after the submission of the Mumtaz Broadcasting Committee Report, its recommendation was partially implemented when Radio Pakistan was converted into an autonomous statutory corporation on 20 December 1972. It was given a new name 'Pakistan Broadcasting Corporation' but retained its call sign 'Radio Pakistan.' Its aim was 'to ensure effective operation and growth of broadcasting as a function-oriented public service medium, general improvement in the quality of programmes, speedy implementation of projects and better utilisation of talent'.

It was expected that the Corporation would be released from the undesirable administrative and financial controls that governed its day to day working as a Government Department. It would be able to plan its programmes with greater freedom, attract talented people through competitive terms, layoff incompetent staff, improve service conditions and freely experiment with new techniques to improve its programmes and hence increase its listenership. The incorporation afforded Radio Pakistan the opportunity to reorganize itself on an efficient operation.

Soon after the conversion, the Board of Directors made necessary organizational changes. Remuneration of employees and artists were enhanced. New departments were created for better management. Yet more than a quarter century after it became a Corporation, the success of its objectives is still indeterminate. A brief description of the outcome of incorporation is given in the last chapter.

Staff increased to nearly double in just two years, and this caused discontent within the organization. A management expert, J.K. Rickard, visited Pakistan under the Technical Cooperation Scheme of the Colombo Plan in May-June 1973 to conduct a manpower survey and suggest organizational changes. Another consultant from the BBC, Peter King, also visited PBC for the same purpose. PBC engaged a firm of management consultants (United Consultants Limited) to downsize and streamline its work systems. The study was started in April 1976 and completed in a year. The Consultants submitted their report in 13 volumes. During 1996-97 a team from the Management Services Division, Cabinet Secretariat, made a thorough study of the overall working of the Corporation with a view to the reorganization of PBC.

7

PROGRAMMING

TYPES AND CATEGORIES OF PROGRAMME

Radio Pakistan in its role as major national institution of audio communication has the responsibility of preserving and popularising this medium in the country. The scope of Radio Pakistan's work in this context includes propagating the national consensus, the evolving political consciousness and the norms of social behaviour. The National Sound Archives preserve the folk-lore, oral history and heritage of Pakistan through the medium of sound. The time-honoured traditions and Islamic values unique to this land are projected in the programmes of Pakistan Broadcasting Corporation.

PBC has broadcast four types of radio programmes. First are the national programmes relayed by all Stations in a nation-wide hook-up, in the national language and sometimes in English. Then there are regional and local programmes that are broadcast in the national and provincial languages from local stations or in provincial hook-ups. And finally there are programmes for overseas Pakistanis and the External Services.

At the time of incorporation in December 1972, PBC retained its old call sign 'Radio Pakistan'. At that time there were nine radio stations putting out 137 hours of programmes daily, including news bulletins. With a radiating power of 756 KW, this covered 38 per cent of the country and 77 per cent of the population on medium wave. External Services was broadcast in 15 languages to about 64 countries for 16 hours daily. By June 2000, Radio Pakistan through its 24 stations was on the air for about 365 hours daily in the Home Service, broadcasting in 21 languages and dialects, namely, Urdu, English, Punjabi, Sindhi, Pushto, Balochi, Pothohari, Hindko, Saraiki, Brahvi, Kashmiri, Balti, Shina, Chitrali, Kohistani, Gojri, Gujarati, Pahari,

Hazargi, Broshashki and Wakhi. It was reaching approximately 78 per cent of the country and 96 per cent of the population.

The Stations are located at Islamabad, Rawalpindi I, II and III, Karachi, Lahore, Quetta, Peshawar, Hyderabad, Khairpur, Multan, Bahawalpur, Khuzdar, Turbat, Faisalabad, Skardu, Gilgit, Dera Ismail Khan, Abbottabad, Chitral, Larkana, Sibi, Loralai and Zhob. Of these, Karachi, Lahore, Quetta, Peshawar and Hyderabad have two channels. Islamabad, Karachi and Lahore put out special FM Service, each for 22 hours from 6 a.m. to 4 a.m.

Nearly half the broadcast time is devoted to information and education and the other half to entertainment. With the opening of the Sautul Quran Channel in January 1998 this scheduling was reviewed. The ratio of religious and spoken-word programmes rose slightly over entertainment items.

PBC programmes cover almost all aspects of life, cater to all segments of society, and its programme content appeals to wide audiences with varied interests.

Information programmes normally comprise news items, programmes on political, economic, scientific, cultural, social and religious topics, and programmes designed for special audiences. Educational programmes are generally curriculum-based. Rural development programmes and mass/adult literacy content are included. Entertainment programmes consist of dramas (including serials), sports, and music (live or recorded) ranging from classical, semi-classical, light classical, folk, light (*ghazal, geet*), film songs, qawwali, recitations, pop and instrumental.

The themes in the information programmes are hot topics in national and international current affairs. The aim is to inform, educate and uplift audiences and build social and political awareness. Topics ranged from export promotion and privatization on the national side to global issues of environment pollution and drug trafficking.

Because of the popularity of morning programmes, these are frequently changed. Generally the morning schedule consists of religious programmes, news bulletins in Urdu, English and regional languages, motivational programmes like 'Raushni', political current affairs items, and light entertainment in the national and regional languages. The morning transmission also includes farming tips from major agricultural centres in the country.

The midday transmission generally comprises of music programmes with national and regional language news bulletins and a segment on religion. At major stations the second and the third transmissions are combined. The third or evening transmission includes, besides religious items, daily special audience programmes aimed at farmers and rural audiences or defence personnel, women's and children's programmes. Towards late evening, more serious programmes are broadcast. These consist of plays, features, panel discussions, sports events and radio reports. After 10 p.m., PBC generally features programmes for more sophisticated tastes such as classical and instrumental music and coverage of local cultural and literary functions, coverage of National/Provincial Assemblies and Senate when in session and programmes for night workers and truck drivers.

For variety and general interest, and to stay competitive in the entertainment business, PBC entered into agreements with BBC, VOA and DW during February-June 2000 to rebroadcast their selected programmes for local listeners.

The main programmes broadcast from Radio Pakistan Stations are briefly described below:

Religious Programmes

Before Independence, British authorities wanted to steer clear of controversy between the two main religious communities of the subcontinent. Hence there were few religious programmes. Immediately before Independence, religious programmes consisted of church services from Calcutta and Tiruchirapalli on Sundays, Bible readings from Dhaka, readings from *Bhagwat Geeta* and *Ramayana* from Delhi and Tiruchirapalli, *Shabds* from Lahore and recitation from the Holy Quran from Delhi, Calcutta, Dhaka and Tiruchirapalli.

Lahore, Peshawar and Lucknow had no religious programmes. Thus only one station within Pakistani territory broadcast *Tilawat* and its translation. With the establishment of Pakistan and consequent change in the complexion of its society and spiritual needs, the recitation of the Holy Quran was made a daily feature from all stations of Radio Pakistan. (Radio Pakistan's transmissions always begin with 'Tilawat' and end with a thought-provoking Quranic verse). Tilawat was followed by 'Dars-i-Quran-i-Majeed' by Maulana Ehtishamul Haq Thanvi from Karachi and relayed by Lahore, Peshawar and Rawalpindi. The object of this '*Dars*' was to present interpretations of Quranic verses from authentic sources to avoid sectarian controversy.

Dhaka Station started its transmission with Dars-i-Quran-i-Majeed in Bengali on four days of the week and in Urdu on three days. In the late fifties, Dars was replaced by 'Muariful Quran by Maulana Mufti Muhammad Shafi on Fridays. On 1 July 1961, the 'Dars' was replaced by another programme entitled 'Quran-i-Hakeem Aur Hamari Zindagi' which explained in simple language the application of the principles of Islam in our daily life. This talk was followed by 'Hamd', 'Naat' and the popular 'Raushni', this last being a talk on Islamic moral and social values.

Dars-i-Quran-i-Majeed is broadcast every Friday from all stations. In the late eighties, these morning national hook-up programmes were reviewed. 'Hayya Alal Falah', a one-hour programme was started, which included 'Tilawat and Tarjuma', a talk entitled 'Quran-i-Hakeem aur Hamari Zindagi', a talk on Islamic teachings and moral values entitled 'Raushni'. Also included was a question period on *Sharia* and Islamic teachings and their application to our daily life 'Aap ne Puchha Hai', Tafseer-i-Quran, Dars-i-Hadees, Hamd and Naat. Another programme 'Khazina' on religious teachings is broadcast daily in the national hook-up after the 1 p.m. news bulletin. *Azan-i-Zuhr* as well as '*dua*' is included in this programme. 'Deen-i-Fitrat' is broadcast on Fridays followed by a question-answer session.

Broadcast of *Khutba-i-Juma* (Friday sermon) began December 1998 from major PBC stations including FM 101, along with an Urdu translation. *Dua* in Urdu is broadcast five times a day in the national hook-up. During the months of Muharram, Rabi-ul-Awwal, Shaban, Ramazan, Shawwal and Zilhijj announcement of sighting of the new moon by the Chairman of the Ruet-i-Hilal Committee is broadcast.

In these religious programmes, the principles of Islam and their application to daily life is explained so as to promote Islamic values. Islam is presented as a progressive and dynamic way of life able to meet the challenges of modern life. Radio Pakistan's basic philosophy on religious subjects has been that Islamic piety is not incompatible with worldly progress. These programmes cover major aspects of Islam, its philosophy, traditions and fundamental principles of faith. Some of the subjects dealt with in these talks include: unity of God, concept of prophethood, respect for life, freedom of will, rights of minorities, equality, social order and justice, contributions of Muslims to science, art and culture among other topics.

Apart from these national hook-up programmes, all stations originate daily programmes on the Quran and Sunnah. These include

'Sarchashma-i-Hidayat' from Karachi, 'Sirat-i-Mustaqim' from Lahore, Hyderabad and Rawalpindi, 'Al-Huda' from Islamabad and Faisalabad, 'Nur-i-Hidayat' from Khairpur and 'Shama-i-Hidayat' from Multan. Most stations also carry a programme 'Aaiye Quran Sharif Parhen' which teaches reading the Quran with the correct pronunciation. In the early years this programme was broadcast under the title 'Sabaq-i-Quran-i-Majeed'. In addition Mehfil-i-Husn-i-Qirat, Mehfil-i-Milad and Naatia Musha'ara are broadcast in the evening from all stations on weekly, fortnightly or monthly basis. These mehfils generally have audience participation. Similar programmes are also broadcast in regional languages from all stations. Devotional songs on Islamic teachings are broadcast regularly in all special audience programmes. All stations end their transmission with the programme 'Din Dunya Ki Bhalai' containing verses from the Holy Quran as inspiration at the day's end.

In addition to these, special religious programmes are broadcast especially during the month of *Ramazan-ul-Mubarak*. During Ramazan a Sehri transmission is added, *Jashn-i-Nuzul-i-Quran* from 22 to 29 and a relay of *Shabina* on the Ramazan nights of 27, 28 and 29; the first fifteen days of *Rabi-ul-Awwal* and the first ten days of *Muharram-ul-Haram* are also routinely brandcasted.

Arrangements are also made to broadcast 'Khatam-ul-Quran' and 'Dua' from *Baitullah Sharif, Makkaht-ul-Mukarramah* direct via satellite during *Ramazan-ul-Murbarak*. During the Ramazan of 1965-66 recordings of 'Ibtehal', devotional songs specially obtained from Radio Egypt, were broadcasted from all stations. It was during Ramazan 1419 AH (1999 CE) that Radio Pakistan for the first time relayed 'Salatul Travih' direct from *Baitullah Sharif* from the 1 to 27 Ramazan. The Muharram programmes culminate in 'Majlis-i-Sham-i-Ghariban' broadcast on the eleventh night in the national hook-up.

Special three-day programmes are broadcast on the death anniversaries of the *Khulafa-i-Rashideen*. Anniversaries of Muslim saints and religious personalities are observed. Programmes are broadcast featuring the life of the Holy Prophet (PBUH), the four Caliphs, *Sahaba-i-Karam* (companions of the Holy Prophet), Muslim saints, religious and spiritual personalities.

PBC organizes an annual *Naat Khwani* competition in collaboration with Pakistan Television. There are three categories of *Naat Khwan*—boys under 15, girls under 15, and boys above the age of 15 years. Selection of the participants for this competition is done at district, division and provincial levels. The winners at the provincial level take part in an all-Pakistan final competition that is broadcast live from all PBC Stations and PTV Centres. A 'Muqabla-i-Husn-i-Qirat' (Qirat Competition) is organized in collaboration with the Ministry of Religious Affairs. The annual International Seerat Conference organized during Rabi-ul-Awwal is provided full radio coverage.

Special training programmes for intending Hajis are arranged in Urdu and regional languages from all stations on a weekly basis, beginning immediately after the announcement of the Haj policy by the Government till the last of the special Haj flights. Daily Haj Service in the World Service is broadcast every year for five months for the benefit of Pakistani Hajis in Makkah and Madinah. Commentaries on all Haj ceremonies including *Ghusl-i-Kaaba* and *Khutba-i-Arafat* are relayed direct through the courtesy of the Government of Saudi Arabia. It has been an annual feature since 1972.

Arabic lessons were introduced from all stations on 21 April 1974 and were subsequently repeated several times. The object of these lessons was to help Pakistanis read and understand the Holy Quran and apply its teachings in their daily life. It also aimed to teach simple Arabic and promote fraternity with the rest of the Muslim world. The World Service also carries Arabic lessons in its regular programming.

Naat Writing Competition

Radio Pakistan organized a *naat* writing (*Naat Goee*) contest for the first time during Rabiul Awwal 1420 AH. Poets were requested to submit their compositions. About 500 poets participated in the competition, each sending four or five *naats* so that over 2000 *naats* were received. A subcommittee judged the *naats* on technical merit and awarded five prizes. The selected *naats* were recorded by performers and broadcast in the morning national hook-up programme. The competition was held again in the year 1421 AH (2000 AD).

Sautul Quran Channel

A separate Quranic Channel was started on 27 Ramazan 1418 AH (26 January 1998). This five-hour broadcast is on the air from 7 a.m. to 12 noon from Islamabad and second channels of Karachi, Lahore,

Peshawar, Quetta and Hyderabad (Islamabad Station joins the channel at 0.15 a.m.). From 11 a.m. to 12 noon recitation of the Holy Quran by renowned international *Qurra* is broadcast. This channel is broadcast in continuation of the national hook-up programme 'Hayya Alal Falah' broadcast daily from 5.45 a.m. to 7 a.m. *Tilawat* (reading from the Holy Quran) is followed by translation in Urdu. Some Stations like Quetta have set up a Sautul Quran Library (inaugurated on 23 April 1998) that preserves *Tilawat* by famous *Qaris* and other Quranic programmes.

The date of the launch of the Sautul Quran Channel has a special significance. On this auspicious lunar date, Pakistan gained its independence. The inauguration of the new channel therefore coincided with the countrys Golden Jubilee celebrations. The Holy Quran was revealed in the holy month of Ramazan and this further added to the auspicious occasion.

Selection of Qaris

The selection of Qaris is an important assignment for Radio Pakistan from the very beginning. Regular auditions were held with the help of important religious personalities and scholars of the Arabic language. Only the best Qaris were offered bookings by PBC stations. In 1950, the Government appointed a Committee for the selection of Qaris with a view to standardize these broadcasts. Members of the Selection Committee were Dr Ishtiaq Husain Qureshi (Deputy Minister for Information and Broadcasting); H.E. El Sayyid Abdul Hamid El Khatib (Minister Plenipotentiary for Saudi Arabia); Maulana Ehtishamul Haq Thanvi (Religious Scholar); Ch. Akbar Husain (Custodian of Evacuee Property); Z.A. Bokhari (Controller of Broadcasting); S. Khurshid (Editor 'Al-Bashir'); and K. Haider (Supervisor, Arabic Unit, External Services, Radio Pakistan, Karachi).

Auditions were held for Qaris from all over the country and the unanimous choice was Qari Zahir Qasmi of Karachi who began broadcasting *Tilawat* from 14 August 1950. Qari Abdul Rehman from Lahore was adjudged as the second best performer. Qari Akhlaq Ahmed was the third choice and he was assigned to broadcast from Dhaka. The Government during 1964-65 set up a high power Selection Board for selecting a new panel of Qaris for the daily *Tilawat* programmes.

Religious Programmes for Minorities

Minorities were allotted time for religious broadcasts on Radio Pakistan from the beginning. Readings from '*Shrimad Bhagwat Geeta*' were broadcast from Dhaka and Karachi every week. These were, however, discontinued after the Indian aggression against Pakistan in September 1965. A fortnightly religious programme for the Buddhist community was started from Dhaka on 13 November 1950, and later on from Rajshahi and Chittagong. It consisted of readings from '*Tripitaka*', the Buddhist holy scripture. Readings from the Bible were broadcast from Dhaka on the last Sunday of every month. Major religious festivals of Hindu, Christian, Sikh and Parsi minority communities are observed.

Mumtaz Broadcasting Committee

The Mumtaz Hasan Broadcasting Committee criticized the language of religious broadcasts as being too formal and unfamiliar. It also criticized the themes chosen for inclusion in these programmes and recommended that these be realistic and applicable to the daily life of the common man. It also recommended that Radio Pakistan should organize special spot recorded *Haj* broadcasts.

MUSIC PROGRAMMES

Fielden once said that music must fill about seven-eighth of any broadcasting schedule. In the late thirties, music (including Western music) constituted on an average about three-fourths of the total transmission. Still in his Report of 1940, Fielden said:

> Broadcasting in most countries does not base its policy exclusively on pleasing the majority. It is to the advantage of broadcasting to widen if it can the scope of listeners' taste... Thus we find in most broadcasting organizations, that classical music gains a place which a majority vote would not actually give it and it is undoubtedly true that the taste for classical music has in many countries been considerably strengthened by this policy.

In the Indo-Pakistan subcontinent, the contributions of men like Amir Khusrau, Sultan Husain Sharqi and Wajid Ali Shah, to Indian music has been tremendous and has contributed to its development.

Just before Independence, a committee was appointed to select the verses of poets ranging from Wali Dakni to contemporary poets for music programmes. The Committee consisted of Mahmood Nizami, Ansar Nasri and Hafeez Hoshiarpuri. The list of selected poems was sent to all stations and was updated from time to time.

In the first two years of Radio Pakistan's history, music programmes suffered because of a lack of talent and the management's preoccupation with programmes of national importance. Indian music did not reflect the aspirations of the young Pakistani nation. This music was mainly of two types—classical and film soundtracks. Classical music, notwithstanding the Muslim contribution, was heavily influenced by Hindu mythology and references, and did not relate to the culture or aspirations of the Pakistani nation.

Film music was considered vulgar and offended Pakistani audiences. It was, however, not possible to break away from the Hindu/Indian tradition overnight. Selections and modifications had to be made. *Thumri* and *Dadra* with sensuous and erotic themes were rejected, while *Khayal* for classical music and *Ghazal* as a form of light music were retained. Words of a number of *Raags* and *Khayals* were changed, and words with an alien bias were replaced by more appropriate words. Experimentation with *qawwali* form and folk music were conducted. Poetry was set to music, particularly those with Islamic and nationalistic themes. In the early years, poetry recitations were frequently featured on radio programmes.

In a new trend, poets and composers drew inspiration from the music of other Muslim countries. An attempt to revive interest in theatre music was made by Central Productions and former theatre musicians. Lahore Station broadcast a series of nine musical features under the title 'Shabistan'. These feature programmes presented theatre music through the years of its development. Stations played musical features based on seasons, popular romantic themes and the works of celebrated national poets, operas based on famous *masnavis*, patriotic songs especially written for Radio, and film music.

Music programme content in Dhaka was slightly different. The musical themes were more heavily steeped in Hindu mythological traditions. The Station attempted a revival of folk music particularly *Ma'arfati* and *Murshidi* types which are in many ways akin to the *qawwali* form in West Pakistan. Dhaka Station also broadcast weekly 'music lessons' to create interest in the folk music of East Pakistan. In West Pakistan, music lessons under the title 'Sargam' by the great maestro, Shahid Ahmad Dehlavi, ran for several years. Music Training Cells were set up in 1965-66 at regional stations to train promising artists under the supervision of veteran musicians.

A Directorate of Music was set up at Radio Pakistan Headquarters in the early fifties. It aimed to carry out research in Pakistani and Indian music, as well as music of Muslim countries, to evolve a new kind of national music. This Pakistani music was to be introduced under a short term and long term plan. The short term plan would concentrate on the improvement of existing music programmes which took up about 40 per cent of broadcast time. The long term plan would encourage the development of Pakistani music.

Radio Pakistan was keen to foster Pakistani music that would be a blend of classical music and folk music of different parts of the country. Music from the different regions of the country has its own unique rhythms and styles. This is aptly described in a passage from a book on Radio Pakistan's early years:

> The Pushto *Tappa* and *Lobha*, with their zest and vigour, the Balochi Ballad with its quick tempo overflowing with valour, the Punjabi *Dhola* and *Mahiya* with their lyrical abundance, the Sindhi *Lehro* with its lilting cadence, the Bengali *Bhawayya* and *Bhatiali* with their gentle rise and fall and their noble magic, provide a feast for all lovers of music. Yet there is an underlying unity in the deeper emotional and spiritual experiences in our folk-songs which, emanating from the common fountain-head of Islam, permeates this variety.

Folk music is at once captivating entertainment and a means for cultural integration. Through Radio Pakistan's promotion of folk music, the whole nation became familiar with the folk music of other areas and even acquired an appreciation for it. Among the successful programmes put out by various stations during the first decade after Independence were 'Des Punjab' and 'Chenab Rang' from Lahore, 'Naghmat-i-Pakistan' put out in the national hook-up from Karachi, series 'Shabistan' on theatre music from Lahore and experimental music entitled *Sargam*, *Darbari* and *Aiman*, prepared by Rawalpindi Station and broadcast from Karachi Station in the national hook-up.

All over the world, music is the mainstay of broadcasting and constitutes the major portion of radio entertainment. Music is one of the main considerations in determining the popularity of a broadcasting station. Radio Pakistan despite the ever-increasing pressure of public service requirements, devotes approximately one half of its total transmission time to music. It presents different kinds of music to its listeners which include folk-songs, classical music, light classical music, *ghazals*, *geets*, *kafis*, patriotic songs, *Iqbaliat*, community songs, large variety of *Hamds* and *Naats*, musical features and *geeton bhari kahani* (story with songs) in the national and regional languages.

All stations strive to maintain a balance, determined by listeners' preferences, between classical, folk and light music. Major stations broadcast classical music on daily or weekly basis under the title 'Ahang-i-Khusravi'. The Classical Music Research Cell of PBC organized monthly seminars and concerts in Lahore and presented classical and semi-classical music recordings by old maestros and live performances by their followers. This programme ran for three years from 1991.

Music is an important part of Pakistan's cultural heritage and one of the basic functions of Radio Pakistan is to preserve, promote and popularize national music. Fortunately, music lends itself naturally to presentation on radio, thereby providing Radio Pakistan with a rich treasure of cultural entertainment material. Radio Pakistan has become the largest music-producing agency in the country. Its output is higher than all the other agencies put together. In fact it can be said that only Radio Pakistan kept alive the tradition of classical and folk music in spite of stiff competition from Western Pop music.

PBC's achievements have been many in the development of regional folk music and of devotional and light music. It has been instrumental in creating a new orchestra, producing folk music with modernized orchestra, producing thematic instrumental music that had hitherto been non-existent in the tradition of Pakistani music. Radio's ongoing efforts to discover and encourage new talent have been pivotal in the evolution of Pakistani music. In the absence of any music schools, Radio Pakistan has provided a platform for musical talent.

Music Talent Hunt

All PBC stations have made attempts to discover new music talent. An estimated 1500 auditions were held in the first three years of the country's existence and about 10 per cent were selected. Another 2500 auditions were held in the next four years. In 1961 and again during 1963-64 a Talent Finding and Talent Utilization Board was appointed by the Director General to tour radio stations in both Wings of the country. The Board auditioned music and drama artists for Radio work. It also reassessed music and drama artists employed by Radio Pakistan to recommend increased benefits to those with outstanding performance records.

In January 1999 a scheme was drawn up to scout for music talent in the country. In the first stage ten major stations (Karachi, Lahore, Peshawar, Quetta, Hyderabad, Multan, Rawalpindi, Khairpur, Bahawalpur and Faisalabad) scouted outside the studios at talent shows and events. Selected performers were invited for studio auditions. The best music artist was called the 'Find of the Month of the Station'. This resulted in the discovery of at least twelve new voices at each station in a year or 120 new voices for Radio Pakistan annually. At the end of the year, a selection of 'Find of the Year' was made from the 12 new voices at each station and the winner was awarded a gold medal and Rs.10,000/- in cash. From these ten winners of the 'Find of the Year' award, the best performer was selected for 'Find of the Year of Radio Pakistan' and was awarded a gold medal and a cash prize of Rs. 25,000/-. At the second stage of this scheme, the remaining stations joined the music talent hunt.

From the talent scouted at regional stations, provincial level competitions were held at Lahore, Peshawar, Quetta and Karachi. The final competition was held at Islamabad on 22 July 2000.

National Songs

In the late fifties the Ministry of Information and Broadcasting produced several national songs on the orders of the President of Pakistan. However, the President rejected all of these songs and directed that this assignment be given to Radio Pakistan. Accordingly, Karachi Station recorded six national songs with orchestra back-up by August 1959. The President approved all of them. This set a precedent and Radio Pakistan began a tradition of producing patriotic songs in a chorus form. The seventeen-day September 1965 war with India was the most productive period for national and patriotic songs.

On Independence Day 1999, Radio Pakistan organized a national song competition and poets were invited to send in their unpublished poems. Around 400 poets participated in this contest. A panel of established poets selected the winners. A prize was given to the best poet of each Province and of the Federal Capital including Northern Areas and Azad Kashmir. The competition was organized during the year 2000 as well.

Community Singing

In the sixties a community choral project was undertaken with educational institutions. The CPU (then Transcription Service) produced a series of live performances at educational institutions by school and college choral groups, which were subsequently broadcast from other stations. The programme ran in both Wings of the country.

Broadcast of Film Songs

Since 1965, film songs broadcast on Radio Pakistan had to be scrutinized by a central authority to bring uniformity in the mode of selection. Lists of approved and banned film songs are circulated to all regional stations. Indian film records were replaced by Pakistani film music in 1964.

Raags for Broadcast

The following Raags were approved for broadcast:

Morning
1. *Bhairon*
2. *Aheer Bhairon*
3. *Lalit*
4. *Mian Ki Todi*
5. *Bilas Khani Todi*

6. *Desi Todi*
7. *Gojri Todi*
8. *Jaunpuri Todi*
9. *Shudh Bilawal*
10. *Alyya Bilawal*
11. *Kalangra*
12. *Aasa*
13. *Hindol*
14. *Ram Kali*
15. *Gun Kali*
16. *Bairagi Bhairon*
17. *Nat Bhairon*

Noon
18. *Shudh Sarang*
19. *Gorh Sarang*

Afternoon
20. *Multani*
21. *Madh Wanti*
22. *Pat Deep*
23. *Puria*
24. *Marwa*
25. *Puria Dhanasri*
26. *Purbi*
27. *Bheem Plasi*

Evening and Night
28. *Eman*
29. *Darbari*
30. *Malkauns*
31. *Bageshri*
32. *Rageshri*
33. *Jai Jai Wanti*
34. *Kamod*
35. *Ghara*
36. *Malganji*
37. *Hans Dhan*
38. *Mian ki Malhar*
39. *Gorh Malhar*

40. *Megh*
41. *Bahar*
42. *Basant*
43. *Adana*
44. *Shahana*
45. *Naeki*
46. *Sooha*
47. Kaidara
48. *Chhaya Nat*
49. *Kalavati*
50. *Gaoti Kalyan*
51. *Puria Kalyan*
52. *Hamer Kalyan*
53. *Shudh Kalyan*
54. *Gorakh Kalyan*
55. *Bhopali*
56. *Behag*
57. *Maru Behag*
58. *Shankara*
59. *Durga*
60. *Aanandi*
61. *Abhogi Kanhra*

Iqbaliat

The name Iqbal immediately brings to mind Pakistan's national poet, Allama Iqbal. It was Iqbal who dreamed of a separate nation for Indian Muslims and inspired them to struggle for freedom. His poetry reflects the spirit and character, the faith and aspirations, and the desire for freedom of the Pakistani nation. After Independence, Radio Pakistan began to spread the message of Iqbal. His poetry became an important feature of Radio programmes, as it was felt the inspiration of Iqbal's poetry could sustain the nation at that crucial time.

In the first two days following Independence, besides *milli tarana*, the Peshawar and Lahore Stations played several compositions of Iqbal (see Box 1).

Iqbal's *ghazals* and poems were ideal for *qawwali* programmes and for poetry recitations. Gradually new ideas dawned for the incorporation of his vision and body of work. Special talks, dramatic

> Box 1
>
> Once again the Caravan of Spring is Here ہوا خیمہ زن کاروان بہار
> Men of Vision create New Environs کریں گے اہل نظر تازہ بستیاں آباد
> The Message of Morning Breeze یہ پیام دے گئی ہے مجھے صبح گاہی
> Is It a Miracle or Time Circle? اعجاز ہے کسی کا یا گردشِ زمانہ

and musical features, orchestral compositions and other programmes on his works were produced.

Some outstanding programmes broadcast to spread the message of Iqbal were 'Aalam-i-Saz-o-Soz' (The World of Sound and Agony) by Faiz Ahmad Faiz, 'Mard-i-Momin'(Man of Faith) by Ejaz Batalvi, 'Masjid-i-Qurtaba' by Hameed Naseem, talk series 'Iqbal ka Ek Sher' (A Verse of Iqbal with explanation) by Ishaq Amritsari and Hameed Naseem from Karachi Station and by Maulana Salahuddin Ahmad and Sufi Tabassum from Lahore Station, 'Mukhtalif Zehnon per Iqbal ka Asar' (Impact of Iqbal on Different Minds) by Dr Raziuddin Siddiqi, talk series 'Chiragh-i-Rah' (Street Light/Beacon Lamp) from Rawalpindi Station, poetic rendering in Urdu and Arabic of 'Shikwa Jawab-i-Shikwa' (Complaint, Counter Complaint) in the voices of Umme Kulsum of Egypt and Noor Jahan.

Radio Pakistan has been criticized for 'overdoing' Iqbal. Yet there are many who believe that because of the place he occupies in the heart of the Pakistani nation and his contribution to Islamic thought and culture, it is impossible to 'overdo' him. In fact, his work should be further popularized for local and international audiences.

Poetic Recitations

Pakistanis have always loved poetry and have great respect for poets. Public recitations of poetry, such as *musha'ara* (audience participated poetic sittings), are universally popular. In keeping with this long-standing literary tradition, all stations of Radio Pakistan regularly broadcast *musha'aras* on a weekly, fortnightly or monthly basis, in Urdu and regional languages. The first radio *musha'ara* was broadcast from Bombay Station at the time Z.A. Bokhari was Station Director. It was organized in connection with the anniversary of the poet Dagh

Dehlavi. Disciples of the famous poet took part. Poets of all ages participates in these *musha'aras* and the list comprises thousands.

At one time Karachi Station used to broadcast *musha'aras* every week as many poets had immigrated to Pakistan from India and most of them settled in Karachi. Because of this weekly frequency the standard of poetry in these broadcasts began to slide. Hence the interval between episodes was lengthened.

Apart from *musha'ara* broadcasts, eminent poets are often invited to recite their new compositions. Most of them read their poetry in a sing-song form called '*Tarannum*' while some prefer regular 'prose-reading', commonly known as '*Tahtul Lafz*'.

Poetry recitals on radio help keep alive the interest in both old and current poetic works. Immediately after Independence, poetry recitations became the mainstay of radio programmes. Similarly during periods of austerity and national mourning, poetic recitations have always filled the void of entertainment in radio programmes.

Western Music

Western music programmes were temporarily suspended after Partition. They were later revived to cater to the demand of a section of listeners and also to spur the evolution of Pakistani music. But problems arose in a shortage of local talent, particularly disc jockeys knowledgeable in Western music. Also, the Stations had a chronic shortage of recorded music. Despite these difficulties Dhaka and Karachi Stations received permission to introduce Western music programmes in 1949 and Lahore followed later.

In Karachi, the USIS and BIS libraries had gramophone records that were borrowed for several years. Live programmes were soon introduced and Western music programmes were started at the other stations viz., Peshawar, Lahore and Rawalpindi as well. The schedule of Western music programmes was published in Pakistan Calling. Critics of Western music programmes prevailed and gradually the duration was decreased and eventually the programmes were stopped altogether in the early eighties.

Another hindrance to the broadcast of Western music was that it was costly to obtain. The Performing Arts Society demanded royalty on the basis of population and not on the basis of actual utilization of gramophone records. Later a demand for a revival of Western music programmes was campaigned in the national press, and eventually they were restarted from Islamabad and then from Karachi, Lahore and Rawalpindi Stations.

Classical Music Research Cell

The Classical Music Research Cell was set up under the Ministry of Education in 1974 and was transferred to Radio Pakistan Lahore in 1978. Its functions include, research in all branches of classical music and publishing of its research findings, research on Islamic musical heritage, building up a library of publications on music, photographs and life sketches of musicians of South Asia. It also aims to encourage Muslim vocal and instrumental artists and promote music through the launch of cassettes, tapes, and records. It also holds monthly concerts to promote classical music.

The Cell accumulated 700 books in Urdu, Hindi and English, 200 articles, 200 pictures of musicians and life sketches of 600 musicians. Some of the books are six hundred to seven hundred years old. Detailed research had been done on seven famous schools of music viz. Gwalior, Delhi, Agra, Patiala, Talwandi, Sham Chaurasi and Kerana. The Cell has been holding monthly functions for many years under different titles such as Ahang-i-Khusravi, Mehakti Sham and Mehfil-i-Musiqi. A revised edition of the catalogues of the material available with the Cell has been published.

FM BROADCASTING

In 1937, the inventor of FM Radio, Edwin Armstrong, built an experimental FM Station in New Jersey, USA. Further development was restricted due to a number of factors, such as common AM-FM ownership which resulted in duplication or repetition of programmes, poor advertising revenues, and competition from AM stations and television. It was only in late fifties that FM Radio gained ground and it has been expanding rapidly since then. Its success is due to its inherent physical advantages in terms of superior sound quality, versatility and stability of coverage. It is static-free and is not affected by evening or late night atmospherics. FM stations direct their programmes to special interest or minority groups. It specializes in music, both classical and soft. It has also developed stereo capability, a feature that PBC should avail of to get better sound quality.

In Pakistan, FM transmitters were used from the beginning but only as a studio-transmitter link. It was used for the first time at Karachi when the studios were at the Intelligence School and the transmitter was located 15 miles away at Landhi.

On 8 April 1993, Radio Pakistan formally began FM transmission under the title 'FM Special', at 7.30 a.m. for two hours only and from three stations—Karachi, Lahore and Islamabad. This was the first music channel for PBC. It aimed at providing light entertainment including popular music by the latest solo and group performances and recorded music produced by EMI and the Shalimar Recording and Broadcasting Company and by PBC itself. These programmes took up about 95 per cent of on-air time. Music segments were interspersed with talk shows, listeners' calls, and public service announcements about traffic education and safety, weather conditions and other local messages.

This transmission became an instant craze with listeners and its duration had to be increased to six hours at each station from 7 a.m. to 1 p.m. daily. At Islamabad another two hours were added in the afternoon from 3 p.m. to 5 p.m. This had the effect of re-claiming Pakistani audiences from listening to stations of neighbouring countries, some of whose programmes were not only vulgar but also hostile to Pakistan. The transmission range was confined to the city of location only. However, Islamabad's broadcasts could be heard from Jhelum to Attock and from Muzaffarabad (Azad Kashmir) to Sargodha because its transmitter was located at high altitude at Murree.

On 30 June 1994, this transmission was given to a private party to manage for one year after which it was taken back by PBC. There was no obvious benefit to the service from this private management. The main reasons for the unpopularity of the service were a lack of interest by producers and presenters. Also the broadcast of AM programmes on FM did not work well for the service as this requires different technical planning and presentation. FM transmission needed to be revitalized to appeal to a younger audience as is the case in most countries.

To make the entertainment programmes of FM Special more attractive and effective PBC decided to revitalize these broadcasts. Young creative teams were formed at the three stations, and the process of planning programmes was changed. The result was 'FM 101' launched on 1 October 1998 from the three stations, i.e. Islamabad, Karachi and Lahore. The service was an instantaneous hit and its duration was increased to 11 hours after one month. On 1 January 1999 its duration was further raised to 17 hours a day and live broadcasts were added. During *Ramazan*, *Iftar* Time Special and after *Travih*, Eid Bazar Shows were broadcast direct from shopping centres in all the three cities and this further boosted its popularity ratings.

FM 101 scored another landmark with its live broadcast of '*basant mela*' (spring festival) on its network on the 13 and 14 February 1999. A month later, presenters at Lahore began operating the control panels. Special programmes were also broadcast in connection with the Independence Day. A one hour daily special on national songs was presented during the first fortnight of August 1999. A live programme was broadcast from the Mazar of Quaid-i-Azam on the morning of 14 August. The solar eclipse of 11 August 1999 was covered live in a one hour fifteen minute transmission. Some experts were invited to provide useful information and to answer questions from listeners. The range of Karachi and Islamabad FM 101 was extended to Hyderabad and Peshawar respectively in January 1999.

FM 101's popularity lay in highly innovative and creative programme planning which made ideal use of FM Service and was aimed in particular at the younger generation. It also helped that the station had powerful transmitters, latest equipment and talented presenters. FM 101 Islamabad has been computerized since 1 October 1999. It has a delay machine, universal access number (UAN-111-111-100) and outstanding facilities, with the added advantage of adequate funds.

On 10 February 2000 the transmission from Islamabad was extended up to 4 a.m. so that it is now on air for 22 hours daily. Karachi and Lahore relay the Islamabad transmission for the extended period. The Karachi transmitter is of 5 KW (stereo) while the Lahore and Islamabad transmitters are of 2 KW each. From 23 March 2000 FM 101 broadcasts are available on the motorway from Islamabad to Lahore due to a new FM transmitter mounted at Kalar Kahar.

It is worth pointing out that commercial service programmes from Islamabad, Lahore, Multan, Quetta and Karachi were being broadcast on FM since 1989, and 'Ahang' carried this information for the first time in its issue of April 1989.

In the West, FM is generally understood to be a music channel, but it has other information programmes such as 24 hour weather updates, traffic situation, and other current affairs issues and general interest topics. As a majority of cars are equipped with an FM radio, drivers are helped with road conditions and traffic updates. It can also be helpful to ambulances and emergency relief services. Presenters and programme producers must research to be competent and helpful.

Another benefit of FM transmission in Pakistan is the fact that a neighbouring country is able to hinder transmission from all PBC stations by using about 10 per cent of its transmitting strength at any time, whereas FM transmission is not affected by such an exigency. To cover the whole country through FM service a network of about 52 FM transmitting stations is needed.

PBA FM School Channel

PBA has the technical capability and equipment to initiate training sessions. To use these facilities a two-hour transmission was started in consultation with the principals of schools and colleges of Rawalpindi and Islamabad, on 2 November 1998 from 9 a.m. to 11 a.m. 'Khushbu' an entertainment programme was on for the first hour while FM school channel was on for the second hour. FM school channel was recorded a day earlier with teachers and students of Rawalpindi/Islamabad educational institutions participating. 'Khushbu' was broadcast live. The school channel was rebroadcast at 4 p.m. for the benefit of morning shift students. This transmission lasted one term only.

Private FM Broadcasts

There were reports in the national press in April 1995 that the Government had granted permission to a private party to operate the first-ever nation-wide cable TV system and three FM radio stations. There was some ambiguity about whether this privatisation was legal. FM transmission under the style 'FM-100' was started from Islamabad, Lahore and Karachi. This 24-hour service consists of pop and folk music as well as old favourites. In April-May 2000 there were again rumours that the Government would allow 500 radio stations to be opened by the private sector in the country.

Mumtaz Hasan Broadcasting Committee Recommendations

The Mumtaz Hasan Broadcasting Committee criticized Radio Pakistan's performance in music entertainment citing a lack of discrimination in the choice of light and film music, poorly produced folk music, injudicious selection of lyrics and artists, bad balancing, comparative neglect of classical music, lack of good instrumental music, and excessive itemisation.

The Committee submitted a report recommending measures to improve the quality and selection of music programmes. They suggested a weeding out of unprofessional voices, greater variety in melodies, better programme planning, expert supervision of live broadcasts, preparation of an anthology of lyrics and songs for use on the Radio. They suggested avoiding *ghazals* with inappropriate rhythms and texts of indifferent literary merit. Pre-recording was necessary for professional management of radio programmes.

The Committee noted that classical music had suffered a decline and such classical forms as *thumri, dadra* and *tappa*, had virtually been discontinued. It recommended that they should be reintroduced.

Recommending that an anthology of *ghazals* should be prepared, the Committee hoped that the selection would avoid poetry that has traditionally 'vague and stylized melancholy, world weariness, self-pity, masochism and naïve romanticism'.

The Committee particularly criticized the *qawwalis* presented by Radio Pakistan, and described them as most disappointing. They recommended that patriotic songs be discontinued, except during a war situation when their popularity ran high. (The Committee was appointed in July 1965 and the Report was submitted in April 1967, i.e. about eighteen months after the 1965 War).

PLAYS AND FEATURES

Plays were broadcast on radio from the very beginning. At that time they were of about two to three hours duration and had songs and intervals like stage plays. When Z.A. Bokhari was Station Director at Bombay in the late thirties he tried to shorten the duration of a play to thirty minutes or one hour. This eventually became the standard length of a radio play. After Independence, Syed Imtiaz Ali Taj realizing the limitations of listeners' attention span cut the duration further to 10 to 15 minutes.

When Delhi came on the ai in January 1936 it tried to develop radio plays as distinct from stage plays. At Lahore and Delhi, through the personal interest of A.S. Bokhari and Syed Rashid Ahmad (who succeeded Z.A. Bokhari as DG, Radio Pakistan), a small and distinguished band of radio playwrights emerged which included Syed Imtiaz Ali Taj, Krishan Chander, Rafi Peer, Rajinder Singh Bedi, Saadat Hasan Manto and Upendra Nath Ashk.

Radio drama is an important source of entertainment, but is a difficult programme to produce. A radio play demands concentration from listeners, does not have the visual appeal of the stage or television and is a challenge for playwrights, producers, actors and the audience. The preference for television as an entertainment medium has made the production of successful radio plays a more demanding undertaking. However, Radio Pakistan has risen to the challenge.

After Partition, there was a shortage of drama scripts and playwrights for radio, and radio producers borrowed old pre-Partition plays that were adaptable to the new political and social condition. Plays with an Islamic bias were unknown before Partition, except for those written by dramatists like Agha Hashr. Writers, both new and seasoned were encouraged to seek inspiration from Islamic history and culture and come up with new themes for radio plays. The response was encouraging and scripts were written showing Muslim pride in Islamic social, moral and cultural values. The stories that came to life included those of Muhammad bin Qasim, Tipu Sultan, Tariq bin Ziad, Salahuddin Ayyubi, Anarkali, Diwana (tracing the history of the spread of Islam in India), Diwaren, Qurtaba ki Masjid were among others.

The aspirations of the Muslim people had become for the first time the focus of radio drama inthe subcontinent. Most of the playwrights in the early days looked for inspiration to other countries. As a result the majority of their works were patterned after plays of other countries and were often alien to local situations and thinking. Four writers, Syed Imtiaz Ali Taj, Rafi Peer, Syed Abid Ali Abid and Khwaja Moinuddin, made important contributions to the development of drama in Pakistan. Khwaja Moinuddin made drama his main pursuit. Syed Imtiaz Ali Taj wrote an interesting play 'Only for the Ears' to demonstrate to new writers that radio was different from the built-in stage and open air platform. It was a light satire on the limitations of radio. Other writers who made tangible contributions include Ishrat

Rahmani, Ansar Nasri, Hakeem Ahmad Shuja, Ahmad Nadeem Qasmi, Intizar Husain, Nasrullah Khan, Saleem Ahmad, Kamal Ahmad Rizvi, Shamsuddin Butt, Agha Nasir, Abdul Majid, Syed Ahmad Riffat, Shama Pervez, Hasina Moin and Bano Qudsia.

Plays and feature programmes based on romantic folk tales unravelled the rich folklore of Pakistan. The dramatized versions of these folk-tales became popular. Translations and adaptations of old stage plays, both foreign and local, filled the gap of available material and themes. Greek plays were easy to adapt and had no copyright considerations. Some English, Russian, American and French novels and short stories were also adapted. Plays available in regional languages were translated into the national language and vice versa for broadcast purposes. The Stations would exchange. This helped to promote national integration through understanding and appreciation of diverse indigenous cultures.

Playwriting competitions were held but the results were disappointing. A training course for radio writers was organized in 1950 at the Headquarters of Radio Pakistan in the expectation that it might help discover new talent. Drama festivals were organized at stations to provide encouragement and inspiration to new writers. Karachi Station held the first drama festival in 1955. Dramatic serials based on Urdu classics like *Bagh-o-Bahar* and *Taubat-un-Nosuh* were undertaken for the first time.

Special interest or thematic plays were also written. An outstanding example of this type is 'The Eyes' (*Ankhain*) by Rafi Peer which was written for the 'Anti-Blindness Week'. Series of plays were written portraying the political, social and cultural achievements of the Muslim people in the days of Islamic glory. Committees of experts were set up in the late sixties at all stations to ensure a regular inflow of drama scripts on different themes. These Committees also undertook a review of national and regional literature for the purpose of selecting material for adaptation into plays.

In the sixties stage plays were broadcast live. For many years music concerts used to be relayed direct from the Al-Hamra in Lahore and The Arts Council in Karachi. The stage play was relayed for the first time in 1962 at Karachi. These plays were specifically written to suit the stage and radio presentations.

In addition to full-length plays every station of Radio Pakistan broadcasts humorous skits on various social themes that have become immensely popular with the listeners. The radio drama has been used as a vehicle of social comment on the evils of bribery, nepotism, social injustice, ostentatious living, etc. Themes advocating child welfare, women's issues and patriotic duty were encouraged. While half the plays deal with social and moral issues, the rest are dramatization of episodes from Islamic history, cultural heritage and folklore.

Drama serials have never been broadcast on a regular basis, however, whenever a serial was run it received rave reviews. 'Nilofer', 'Musafir Manzilen aur Raste', 'Emergency Ward', 'Anwar' and 'Afshan' can be cited as outstanding examples of drama series.

Radio drama used to be a popular feature but went into decline for about sixteen years. Finally, Karachi Station revived the tradition of drama festivals and organized *Jashn-i-Tamseel* in November 1991. Shortly after, the Station broadcast the recordings of fourteen hit plays during June 1992. Other stations followed Karachi's lead and Peshawar Station organized a drama week of Pushto plays in the last week of August 1992. Lahore, Multan and Rawalpindi stations also observed drama festivals, which were a great success and received favourable press reviews. This proved that dramas were successful not only for entertainment but also for projection of public service themes and campaigns.

During 1993, a special series of plays 'Sachi Kahanian' (True Stories) containing thirteen episodes on narcotics control was produced and broadcast from all capital stations in cooperation with the Integrated Drug Demand Reduction Project of UNDP. Drama weeks are now a regular feature at major stations.

During 2001, a drama serial 'Dukh Sukh Apne' (Our Joys and Sorrows) was launched to educate people on reproductive health and other related issues. This UNFPA funded project was to continue for two years, and was supposed to feature 104 episodes. The episodes were to be translated into Punjabi, Sindhi, Pushto and Balochi. The first episode of the Urdu version went on air on 11 February 2001 in the national hook-up. The broadcast in regional languages was started in June 2001.

A seminar was organized in Karachi on 26 May 1992, to revive interest in radio plays and to improve their quality and content. Participants included experienced broadcasters, writers and radio drama enthusiasts. Similar seminars were organized at other radio stations also with interesting results.

Lahore Station observed 1997 as 'drama year' and featured plays by established playwrights and dramas written in connection with the Golden Jubilee of Pakistan. Classic plays were broadcast in the second quarter of 2000. PBC started airing old radio plays, including classic plays of the fifties and the sixties, every Saturday simultaneously from eleven major stations including Islamabad, Karachi, Lahore, Peshawar, Hyderabad and Khairpur. The plays selected were mainly on social issues including human rights, narcotics, bonded labour and women's rights.

FEATURES

A radio feature is a unique format in broadcasting. It was unheard of in pre-radio days. A feature presents information in detail with a particular viewpoint similar to magazine and documentary. The main difference between a feature and a magazine is that the latter tends to treat two or more subjects within one programme whereas the radio feature restricts itself to one subject. The difference between a feature and a documentary is of approach. A feature allows much more subjective treatment than a documentary. Radio feature requires research like a documentary.

It can be said that a feature is a 'creative interpretation of reality' and is gainfully used in projecting any one theme or idea with an imaginative mixture of music, poetry, narration, sound effects, dramatization and actuality. When produced well, a feature can leave a lasting impact on the listener. Unfortunately, features have been misunderstood to be narration in two or more voices. Features have been mostly on developmental or social welfare themes.

The lack of new radio plays in the early days was made up with feature programmes. Some outstanding and popular examples of feature programmes are 'Pakistan Hamara Hai', 'Zindabad Pakistan', 'Istiqlal-i-Pakistan', 'Terah Sau Sal', 'Jang-i-Azadi', 'Alhamra', 'Baghair Unwan Ke', 'Hota Hai Jada Paima', 'Karenge Ahl-i-Nazar Taza Bastian Abad', 'Dhunde Hai Us Mughanni-i-Aatish Nafas Ko Ji', 'Jhelum', 'Qaziji', 'Hamid Mian Ke Han' (it ran for over 30 years), 'Dekhta Chala Gaya', Danishkada, serials 'Shama' and 'Manzil', series 'Inspector Shahbaz Khan', 'Muft Ka Jhagra', 'Talqeen Shah', 'Juma Juma Aath Din', 'Hamari Kahani Terikh ki Zabani',

'Nishan-i-Rah', 'Delhi Ka Akhri Badshah', 'Chacha Umardraz', among others.

Besides nation-building themes these features dealt with a myriad of issues on history, culture, rise and fall of Muslims, the freedom struggle, reconstruction and rehabilitation, crime detection, reformation, and common social issues.

The Mumtaz Hasan Broadcasting Committee criticized radio drama for restricting itself to a limited number of themes. The problem was that a certain code for acceptability has evolved over the years that is inhibiting for the creative writer. The Committee recommended that Radio Pakistan should make full use of outside production talent if necessary on a contract basis, and it should pre-record its dramas. The Committee also recommended pay increases to dramatists and artists.

TALKS AND DISCUSSIONS

Talks and discussions are the main instrument for providing information to listeners and educating them on life's important issues. A good 'talk' is a friendly chat built around one particular subject of general interest with examples of personal experience. The main ingredients of a talk are unity (sticking to the main idea from beginning to end), clarity (short and simple sentences that can be easily understood) and emphasis (correct expression, pauses and stresses). The duration of a talk normally varies from two to five minutes. It can be of a longer duration also, but the aim is to hold the interest of the listener.

Talks are of two kinds—those that are planned in advance in the form of quarterly schedules and others called 'topical' talks which are arranged at short notice and focus on some important issue or event. Preparation of the advance quarterly talks schedule is an important undertaking at each station. Producers do a lot of research in consultation with scholars and field experts to come up with ideas and then develop them. Expert advice becomes necessary in working out a series of talks on specialized subjects. Some ideas take the shape of a 'series' of talks and cover different aspects of a theme, while others are worked out as 'individual' or 'stray' talks.

These schedules are discussed thoroughly by the heads of stations before being submitted to Headquarters, where they are again scrutinized. One or two of the series are selected as national hook-up and/or inter-station series depending upon their importance and relevance. Great care is taken in the range and variety of subjects and the choice of speakers and at all stages of planning, production and presentation. The language of the talks has to be easily understood, conversational, informal and personalized rather than weighty and academic.

A discussion programme provides a forum for the exchange of ideas. In a discussion more than two (usually three or four) persons discuss a topic presenting more than one point of view. The discussion may be based on a simple or controversial idea. The discussion can be serious or light-hearted but the aim is to be thought-provoking. A discussion is always unscripted. There is a chairman/moderator to lead the discussion in the right direction. He also sums up the discussion, but he must never conclude with a verdict.

The main aim of talks and discussion programmes at PBC are to build awareness and pride in our cultural heritage, educate listeners about the civic responsibilities and to spur cultural integration and tolerance on a national, global and Islamic-nations level.

Series of talks on Islamic culture have been broadcast. Other topics include readings from Islamic history and talks on the administrative system of Muslim governments of the past and of the Islamic countries of today, with particular stress on their democratic foundations. Islam is discussed as a code of life and its concept of morality is brought out in other series. Life and works of Muslim historians, travellers, philosophers, scientists, military men, reformers, scholars, intellectuals, saints and other prominent personalities have formed the subject of several broadcasts. Talk shows on the spread of Islam to various regions of the world and the important centres of Muslim culture have been broadcast.

A series of talks was produced reviewing the struggle of Muslims of the subcontinent for the creation of Pakistan and the economic development of the country. Another important series was 'The Struggle for Pakistan' by M.H. Syed broadcast from Karachi. This series traced the history of the political endeavour, culminating in the establishment of a new independent Muslim state in the Asian continent.

Series of talks on the life and works of Pakistani poets, authors, artists and other creative personalities are broadcast by local stations as well as by other broadcasting units. These talks aim at fostering an appreciation of Pakistan's cultural diversity. It was these talks that made legends such as Khushhal Khan Khatak, Rehman Baba, Baba Ghulam Farid, Bulhe Shah, Madho Lal Husain, Shah Abdul Latif Bhitai, Sachal Sarmast, Mast Twakkali and Jam Durak known to the average Pakistani.

Similarly talks on the life and history of Pakistan's towns and villages have been regularly exchanged among the stations. National programmes that are centralized at Islamabad (formerly at Karachi), are relayed by all stations.

Several series on science subjects have also been started. During 1991 Lahore Station in collaboration with Science International organized a series of seminars on science and technology. In this year-long programme two types of research papers were presented, one for scientific experts and the other for the general public. Talks interpreting science for the common man, talks on modern psychology, present day economic problems, current affairs, book reviews etc. also constitute a regular feature of the spoken-word broadcasts of all stations. There have also been discussion programmes on highly controversial problems of the day. Such topics include nationalization of private property, Pakistan's attitude towards the superpowers, membership of defence pacts, feudal system of land ownership and others. In a lighter vein, series of talks highlighting day to day events in a city proved very successful.

The subjects of talks range from teaching good citizenship to the latest advances in scientific research and higher aesthetic and spiritual values. A concerted effort has been made to replace straight talks by illustrated talks, panel discussions and interviews, which according to audience research carried out by Radio Pakistan, have greater listener appeal. Brains Trust and *Danishkada* in Urdu are popular audience interactive programmes.

On the subject of talks, mention may be made of the President's first-of-the-month broadcasts introduced from October 1963. In fact Liaquat Ali Khan initiated the first-of-the-month broadcast. Prime Minister Chaudhry Muhammad Ali adhered to this tradition starting in July 1953 to the end of his office. This was done to give a panoramic view of the development of different activities in the country, the Government's policy on current national and international events as well as Pakistan's contribution to international peace and progress. Thereafter middle of the month broadcasts by the Governors of East Pakistan and West Pakistan were also started in 1963-4.

MEMORIAL LECTURES

Memorial lectures have been organized by Radio Pakistan to commemorate the name of the Nation's Poet, Allama Iqbal. The series entitled 'Iqbal Memorial Lectures' was initiated on the twenty-seventh death anniversary of the poet in 1965. The Minister for Information and Broadcasting inaugurated the project. A series of English lectures by Dr Abdus Salam, who at that time was Chief Scientific Adviser to the President and Director of the International Centre of Theoretical Physics, Trieste, Italy, was a prestigious project. The subject was 'Symmetry, Matter and Energy in the Universe', and Dr Abdus Salam dealt with the concept of fundamental particles giving a revolutionary interpretation of nature and structure of matter. These lectures representing the quest of the physicist for the ultimate Reality, were translated from English into Urdu and Bengali and published in a booklet.

The Quaid-i-Azam Memorial Lectures were broadcast in 1969-70 on the subject of 'Ideology of Pakistan'. Each lecture was delivered by a different scholar and was followed by a question period. A list of distinguished academics who delivered the Quaid-i-Azam Memorial Lectures:

1. Dr Ishtiaq Husain Qureshi,
   Vice-Chancellor,
   Karachi University.
2. Dr Javed Iqbal,
   Scholar/Intellectual, Lahore.
3. Syed Sajjad Husain
   Vice Chancellor,
   Rajshahi University.
4. Mr Muhammad Ali,
   Ex-Vice Chancellor,
   Peshawar University.
5. Mr Justice Abu Sayeed Chowdhary,
   Vice Chancellor,
   Dhaka University.

Memorial lectures projecting important personalities and subjects are arranged by different stations on relevant occasions.

Documentaries and Outside Broadcasts

Outside broadcasts (OB) are intended to air short-lived events happening outside the broadcasting studios. Radio documentaries and actuality programmes on the other hand are based on subjects such as institutions, development projects and historical places. The interest in an OB mainly lies in its topicality while in the case of a documentary or an actuality programme it rests on the ability of the programmer to focus on points of interest or historical significance.

The documentary format is used to deal with one particular subject of interest based on research and oral evidence, and seeks to inform and educate. Its basic structure is content (information) and treatment (presentation). A documentary combines quite a number of radio presentation forms such as news, talk, discussion, interview, drama, music, sound effects, silence and actuality recording. It requires expert handling of the material used both in script and production.

Despite numerous handicaps in staffing and equipment, Radio Pakistan has always risen to the occasion whenever there is an event of national importance. On the very first day of its existence, it broadcast the Independence Day celebrations.

The number of events covered through OBs runs into thousands every year, ranging from public *musha'aras*, sports events (national and international), military functions, university convocations, national and international conferences, educational and cultural events, Federal and Provincial budget presentations, press conferences and official tours at home or abroad and crisis events among others. OB teams visit villages in the interior and tribal areas to prepare actuality programmes for rural listeners.

The main object of documentaries is to highlight various completed projects of socio-economic development. Documentary programmes that acquaint listeners with national and global issues of the environment, pollution, wildlife preservation, etc. are also broadcast. National issues like development projects, agriculture, education, etc. are presented. During 1964-65 Karachi Station broadcast a series of documentary programmes on major industries, preceded by talks by eminent industrialists. These were later published as a book.

Other events that for whatever reason could not be covered direct are recorded and later on edited in the studios and presented in the form of a 'round-up'. Actuality programmes, round-ups, radio reports, eyewitness accounts and documentary programmes are popular with listeners and are considered as the greatest test of the professional efficiency of a Radio Producer.

Budget Coverage

All stations of PBC cover the Federal and Provincial budgets in the national and provincial languages. Coverage runs over a period of a month with pre-and post budget discussions. The budget speech of the Federal Finance Minister is relayed direct from the National Assembly Hall in a special national hook-up. The reactions of cross-sections of the people and salient features of the national budget are broadcast in the national and regional language programmes and news bulletins. Discussions are held on the economic and fiscal policies of the Government and the significance and salient features of the budget. Coverage and commentaries on the Economic Survey issued before budget and on the post-budget press conference are also featured.

The budget speeches of the Provincial Finance Ministers are also relayed by the respective Stations, direct from the Provincial Assemblies, followed by salient features in the national and regional languages.

Senate, National/Provincial Assemblies Coverage

Proceedings of the Senate and the National Assembly, whenever in session, are covered in a special 15-minute national hook-up programme. Some speeches are relayed direct, such as the President's speech to the first joint session at the start of the Parliamentary year, budget speech of the Federal Finance Minister, address of a visiting foreign dignitary to Parliament, oath-taking ceremony of a newly elected House, elections of Speaker, Deputy Speaker and Leader of the House and vote of confidence.

Similarly, all stations cover the proceedings of their respective Provincial Assemblies in Urdu and regional languages. Elections of Speakers, Deputy Speakers and Leaders of the House and votes of confidence are normally presented in the form of radio reports.

President's/Prime Minister's Visits Abroad

Visits abroad by the President and Prime Minister receive coverage by radio reporters. Special broadcasting teams accompany the President or the Prime Minister on these trips to ensure prompt and accurate coverage of the tours. PBC also supplies specially recorded programmes in advance to the broadcasting organizations of the host countries. These may include personality profiles of the President/ Prime Minister, documentaries on the country, as well as Pakistani music.

A curtain-raiser programme is broadcast in the national hook-up on the eve of the delegation's departure giving basic information about the country to be visited, its historical, geographical, social and cultural background, its relations with Pakistan and the significance of the proposed visit. Thereafter daily radio reports are broadcast containing eyewitness accounts of the arrival, important speeches at State functions, press conferences and special interviews with dignitaries as well as a 'sound picture' of each day's engagements, and reports or round-ups on the return. If considered necessary, the President or the Prime Minister's speeches are relayed direct from the foreign country.

Similarly, visits of foreign dignitaries are covered with documentary programmes on the country concerned, and live or recorded radio reports on the visit. Voice-cast and studio facilities are provided to the foreign press.

Sports

PBC has extensive coverage of sports activities that has brought appreciation and criticism from the public. Sports lovers appreciate it and praise PBC for promoting sports in the country through its running commentaries especially on hockey and cricket matches. However, there is also criticism of too much sports coverage, particularly cricket.

Sports events at home and abroad in which Pakistan participates is given live coverage or report summaries in the national and regional news bulletins. A programme 'Aalami Sports Round-up' is aired daily at night in the national hook-up giving brief reports on sports events around the world. The morning national hook-up programme 'Subh-i-Pakistan' also carries a sports report. In addition, all PBC stations in their regular sports magazine programmes broadcast talks and discussions on various aspects of the games.

Local stations cover major sports activities in the news and through running commentaries, eyewitness accounts, radio reports and interviews.

Programmes on Science and Technology

Programmes on science and technology have been broadcast from the very beginning, but with less emphasis than these subjects deserve. These programmes include talks and discussions on science, health, hygiene and nutrition and special audience programmes particularly for women and farmers. A number of science-related series were initiated from different stations in 1948-49, such as 'Science ke Maarke' (Achievements of Science), 'Science aur Ijtemai Zindagi'(Science and Society), Science Magazine, Health Tips, Radio Doctor, Family Doctor, Nutrition etc.

In 1950, the Lahore Station broadcast a series on 'Muslim Scientists' describing the contributions of Muslims in various branches of science. Research in Pakistan in different fields is also a regular feature of programmes. These include work being done in such institutions as Pakistan Atomic Energy Commission (PAEC), PINSTECH, A Q Laboratories, PCSIR, NARC, DESTO, Pakistan Science Foundation, Pakistan Engineering Council, as well as at Universities. Important national and international science conferences are covered in radio programmes. These include Pakistan Science Conference, Colloquium on 'Science at Mid-Century', PCSIR, atomic explosion, among others. Interviews of renowned scientists are broadcast as and when possible. The interview with Sir Alexander Fleming, discoverer of Penicillin, was broadcast on 16 April 1951. Other scientists featured include Dr Salimuzzaman Siddiqi, Dr Raziuddin Siddiqi, Dr I.H. Usmani, Dr Attaur Rehman, Dr Abdus Salam, Dr Ashfaq Ahmad, Dr Samar Mubarakmand and Dr Abdul Qadeer Khan.

The Hyderabad Station has played a very important role in the propagation of scientific information particularly in the field of genetics. From 1956-57 the Station featured programmes in Urdu and Sindhi. The translation in Sindhi was broadcast from Khairpur Station for the benefit of local agriculturists and farmers. Whenever some national or international conference was held its proceedings were given suitable coverage and interviews of delegates were aired. For example, in Pakistan the first international symposium on 'New Genetical Approaches to Crop Improvement' was held in February 1982 at the Atomic Energy Agricultural Research Centre, Tando Jam. Interviews of participating scientists from Germany, Greece, Japan, England, Sweden, Italy, Australia and USA were broadcast from Hyderabad Station.

The second international symposium was held in Karachi in 1992 and the next in 1997 in Tando Jam. The latest scientific information presented at the symposia is disseminated in radio programmes. An example is the case where high-yield seeds were developed and the advantages of their use were publicized in a radio programme. Farmers were encouraged to buy the new seeds and benefit from higher per acre yield and bumper crops.

Similarly, Radio Pakistan Lahore in collaboration with Science International organized a series of seminars on science and technology. Besides dwelling on the latest technology, the seminars presented the works of those scientists who had been published in foreign science journals but were relatively unknown within the country. In this year-long programme two types of research papers were presented, those aimed at the general public and those for the scientific community. The first seminar in the series was held on 1 February 1991.

The title of the Iqbal Memorial Lectures, 1965 was 'Symmetry, Matter and Energy in the Universe' representing the quest for the ultimate Reality.

In the first few decades after Independence, '*Science ki Dunya*' (The World of Science) 'Science Magazine', '*Science ki Baten*' had become regular weekly features at all stations. Science-related questions were invariably included in all quiz programmes to foster interest in science subjects and research. Two regular series of talks are broadcast every week from almost all stations for the last three decades. These series, aired in collaboration with the USIS, are entitled 'Science ki Dunya' and 'Mukhtasar Mukhtasar' (Snippets). These give useful information on the latest research being done worldwide in various branches of science.

Literary and Cultural Programmes

The educated class in Pakistan has a great appreciation for literature and the Arts. Literature is a popular choice of study at college and university levels, and many students maintain literary pursuits after graduation. Poets and poetry are held in high esteem in the country. PBC has given literary programmes regular air-time in national and regional languages from all stations on a weekly or fortnightly basis.

The weekly broadcast of literary magazine has become a permanent feature in which modern literary trends are discussed. Another regular feature are the *musha'aras* (Poetry recitation sittings) usually broadcast at short intervals. A *musha'ara* is a gathering in which poets recite their own verses composed specially for the occasion. The poems are either of common rhyme-scheme or a common theme which lends a competitive flavour to the performance. Each poet presents his work direct to the audience. The gathering conveys its approval by applause and disapproval by silence or other signs. These functions can get emotionally charged depending on the presentations. *Musha'aras* are popular though expensive to produce programmes.

Important cultural and social events are also covered through radio reports. All stations of Radio Pakistan broadcast daily radio reports under the title 'Shaharnama/Khabarnama', projecting social and cultural functions held in the city. In 1964-65 Lahore, Rawalpindi and Peshawar Stations broadcast a joint monthly production of cultural features. Book reviews are a regular feature at all stations. Previously only books of fiction and poetry used to be reviewed. In 1975, the scope was widened to include books published in Pakistan in such fields as economics, science and agriculture. New and young writers are afforded opportunity to express themselves along with the well-established and well-known litterateurs.

Some of the topics covered in literary programmes are: tradition and innovation in literature, Islamic mysticism and Pakistani literature, Islamic system as the basis of national character, modern literary trends and collective consciousness and essential factors of Pakistani culture.

Anniversaries and Festivals

In undivided India, a list of anniversaries and festivals was given to all stations of AIR. In this list programmes for religious communities and cultural groups were allocated. Some of these groups had to campaign for their share of air-time.

After Independence, the change in the complexion of society in Pakistan called for a change in the list of religious programming. Due to the overwhelming majority of Muslims in the country, their spiritual needs inevitably received greater reflection in the revised list. At the same time all the important festivals of the minority communities were included. The new list recognizes prominent men in the history of Islam, leaders of Indian Muslims who had struggled for independence, literati, scholars, saints, scientists, military generals, politicians and other leaders of repute. Important festivals of the minorities (Hindus, Scheduled Castes, Parsis, Christians, Sikhs and Buddhists) are also observed. Special cause dedications announced by the United Nations, its specialized agencies and other international organizations are also observed by PBC. National days of friendly countries are observed on reciprocal basis.

Full-day programmes are broadcast on special occasions such as Pakistan Day, Independence Day, birth and death anniversaries of the Quaid-i-Azam, death anniversaries of Allama Iqbal and Liaquat Ali Khan, 6 September, *Eid-ul-Fitr, Eid-ul-Azha, Milad-un-Nabi* among others.

Special programmes are also featured during the first fifteen days of *Rabi-ul-Awwal*, first ten days of *Muharram* and the whole month of *Ramazan-ul-Mubarak*, including a special *Sahri* transmission from all stations of PBC.

For the first time in the history of Radio Pakistan messages of the Heads of State of Turkey and Iran were broadcast on the Quaid-i-Azam's death anniversary on 11 September 1964, paying glorious tributes to the Father of the Nation.

RURAL AND FARM BROADCASTS

Rural Broadcasts

As nearly 70 per cent of the country's population lives in villages many of PBC programmes are rural-oriented. It is particularly in this instance that Radio Pakistan has always been conscious of its role as an instrument of change, education and information, besides entertainment. Radio has become a knowledgeable guide for rural listeners and their specific needs and problems are reflected in specialized programmes.

Programmes for the rural population were included from the early days of broadcasting in the subcontinent. For example, when Peshawar Station was set up on an experimental basis in 1935, it was primarily meant to serve community listening at the *Hujra*, or at *utaq, chaupal, panchayat ghar* or community centre. Even today special programmes are planned and produced from the point of view of community listening and are broadcast in regional languages. The presenters are specially chosen for their easy country style and wit. These programmes are generally broadcast in the evening which is prime time listening for the villagers. The content is usually music (both folk and film), rural news bulletins, weather reports, market reports, short plays, skits and talks.

The talks mostly deal with religion, day to day problems of the villagers, elementary science, cottage industries, hints on better farming methods, education, health and savings. Plays, skits and features deal with village issues, such as the evils of illiteracy, wasteful expenditure, personal debt, litigation, marriage relationships and vendettas.

The special composite programmes directed at rural audiences are broadcast from all stations of PBC except Islamabad for 30 to 135 minutes daily in regional languages and dialects. The main object of these programmes is to inform and educate village folk so that they can contribute gainfully to the socio-economic and cultural life of the country.

Peshawar Station was broadcasting a special programme for tribal listeners similar to the rural programme. As far as language, level of education and social progress and entertainment needs of the two groups are concerned, they are very similar. Eventually the two programmes were merged together. This combined programme aims at eradication of evils of superstition, ignorance, litigation, disease and revenge still prevalent among certain classes, and to help people raise their economic and educational standards. There are indications that the programme is having a strong positive effect on the rural and tribal population of NWFP.

This programme required careful planning to avoid confrontation with groups of differing religious views across the international border. After January 1950, this programme was used for countering the propaganda launched by Radio Kabul against Pakistan and the Tribesmen from the time of Partition.

As there was no Radio Station in the Provinces of Balochistan and Sindh except at Karachi, the latter was required to meet the broadcasting needs of both provinces. Programmes were started in Sindhi (from the very day of inauguration of Karachi Station, i.e. 14 August 1948) and in Balochi (from 25 December 1949). In the absence of a dedicated rural programme these catered to the rural needs as well. This arrangement continued till the opening of Hyderabad and Quetta Stations. Programmes in Sindhi were reintroduced from Karachi Station from 1 July 1970.

Farm Forum

Like rural broadcasts, programmes for the benefit of farmers also began from the early days. However, the first formal programme directed at farmers was started from Peshawar Station on 5 January 1960 on a weekly basis. Regular farm forum programmes were introduced from 1 December 1966 from five Stations (Lahore, Hyderabad and Peshawar in West Pakistan and two in East Pakistan) on an experimental basis. Following its success, the duration of these broadcasts was increased and Quetta, Rawalpindi and Multan Stations also started daily farm broadcasts for about thirty minutes to an hour.

Radio is considered the most effective communication medium for farm broadcasting because transmission covers vast areas and reaches remote rural areas. Expert advice can be disseminated in a variety of interesting ways. Farmers are able to carry on with their work while listening to the programmes. Most compellingly it is the ideal medium for educating the illiterate.

Farm broadcasts are arranged with the active help and close cooperation of the Federal Ministry of Agriculture, Provincial Agriculture Departments, Pakistan Agricultural Research Council and the Agricultural Development Bank of Pakistan. Organizations providing facilities and services are also associated.

The aims of farm broadcasts include:
a) To focus attention on the day to day problems of the farmers and to advise them on preparation of soil, application of fertilizers, seeds, sowing patterns, watering the crops, weeding out operations, use of insecticides, harvesting, marketing and storage.
b) To motivate farmers to use improved means of agriculture.

c) To acquaint the farmer with the proper methods of livestock rearing, poultry farming, fish farming, dairy farming, agriculture, afforestation, sericulture, operation and maintenance of farm machinery.
d) To inform the farming community about the activities of the agriculture departments and allied organizations and agencies for resolving the problems faced by the agriculturists.
e) To motivate the villagers to undertake farming on cooperative basis.
f) To motivate the farmers to make the best use of their time, energy, capabilities and resources, so as to get the maximum yield and ensure their own economic betterment and substantial contribution to the national economy.

With the introduction of the daily composite farm broadcasts the campaign to reach the farmer with expert advice and latest farming information assumed a new dimension. The farm broadcasts were instrumental in changing the age-old farming practices of conservative farmers and spurring the acceptance of new technologies.

Here it may be relevant to quote Syed Munir Hussain, former Director General (1966-69) from his letter of 14 January 1997 to the Director General PBC:

> It is difficult to recount the various services Radio rendered to the people in its coverage of social, cultural, economic, rural, agricultural, industrial, health, education, women and children issues; but one that specially needs to be mentioned is its immense contribution in propagating the adoption of Mexi Pak wheat seed in West Pakistan and Irri Rice in East Pakistan for raising the yield of these crops. Their extensive adoption heralded the coming in of the Green Revolution in agriculture in the country. It was not easy to break the age old notions and practices of the farmers but the manner in which Radio embarked on this venture by addressing the agricultural community through extremely well-conceived actuality programmes, no other field organization could have accomplished this, in the short time we witnessed the use of these improved variety of seeds.

Farm broadcasting is a complex, multi-dimensional and multi-disciplinary activity. It involves meticulous planning as it has to integrate the communication needs of all the agencies working in the field to assist the farmer. This requires the association of the agriculture extension worker, the mobile credit officer, the veterinarian and the livestock breeder, soil scientist, poultry and fish farming advisers and representatives of agencies providing seeds, fertilizers and agricultural machinery.

For successful farm broadcasting objectives had to be defined and programme schedules had to be planned for maximum advantage to the listeners. Producers had to establish institutional linkages for flow and dissemination of information. Also required is training of communicators and personnel, organizing of farm listening groups with the help of extension workers, conducting audience research, recording and documentation of field interviews, and hiring rural correspondents. Producers must devise a different programme for each agro-ecological zone, highlighting the package of agriculture technology devised by the Pakistan Agriculture Research Council from season to season and from crop to crop.

The programmes are broadcast in local dialects for better understanding using stock characters well-versed in the local idiom. The easy rapport, cultural affinity and humour of the comperes contribute to the popularity of the broadcast.

Through these programmes information and advice is given to the farmers daily in an easy and understandable form, interspersed with songs, features and witty dialogues, interviews with progressive farmers, talks and discussions by experts, question-answer sessions, spot announcements, and agricultural news.

During the year 1968 the Department of Agriculture undertook the task of setting up Farm Forums, or listening clubs, in selected villages on an experimental basis. These forums were village radio clubs organized with the object of ensuring better listening, bringing in feedback to the programme planners and producers and providing a base for the evaluation of the programme. The idea is to have accessible farm forums on the pattern originated in Canada in 1941 and followed in Italy, Japan, New Zealand, India and other countries. Many Farm Forum Clubs are at present functioning regularly with the transistor sets supplied to them by the Bureau of Agricultural Information.

All Stations of PBC except Islamabad broadcast 30 to 50 minutes programmes daily in local dialects to instruct and motivate farmers to adopt modern, progressive and scientific methods of farming. Listeners are also kept informed of the Government's activities in the field of agriculture, e.g. land reforms, revenue remission reforms and their implications, credit availability and measures to increase yield per acre. In addition to the evening farm forum programmes, a 15 to 30 minute programme for farmers runs in the morning or midday transmission from Lahore, Rawalpindi, Multan and Hyderabad Stations. The average time devoted to this programme is about eight hours per day, and the duration of rural and farm broadcasts is about twenty hours a day.

These two programmes directed at rural listeners have been effectively used for the uplift and development of rural areas. Stations have acted as intermediaries between the villages and concerned departments or agencies for the solution of problems faced by the villagers. An example in the late seventies was when Hyderabad Station selected a village, Phalkara, which lacked almost all basic civic facilities to initiate development works. It took nearly two years hard work to turn this village into a model community enjoying all civic amenities such as potable water, sewerage system, primary school, dispensary, police station and offices of local administration.

This experiment was so successful that Radio Stations received requests from other villages for similar projects. In October 1997, Multan Station also embarked on such a project and selected a village, Jalalabad Shumali, for the purpose. The main problems tackled with the help of local Administration were setting up a sewerage system, girls school, roads, lanes and culverts. A report was recorded every fortnight in the village to monitor the progress of the project for broadcast in the rural programme of the Station.

The Mumtaz Hasan Broadcasting Committee observed that the programmes for rural listeners were the most popular. It was felt that the work of Union Councils and Union Committees received too much attention in these programmes.

Educational Broadcasts

Radio was once a synonym for entertainment but is now recognized as a useful medium for imparting education. All advanced nations of the world have used this powerful medium in the field of education to their advantage. In Pakistan low literacy rates are a hindrance to economic development. New schools, colleges and universities are being opened and existing institutions expanded to fill in the vast gap. Radio Pakistan also came forward to play its role to help promote the cause of education. Besides general educational programmes, most stations broadcast regular programmes for school and college students drawn up in consultation with educational experts. These are regularly transmitted to an ever-increasing number of listeners and have proved beneficial to the student community.

Educational broadcasts of Radio Pakistan consist of programmes for school, college and university students, programmes of mass literacy and adult education, Teachers' Training and Allama Iqbal Open University Programmes. (This last item is discussed under Commercial Service).

Programmes for School Students

Programmes for school students were among the first broadcasts from Lahore and Dhaka Stations. These programmes were planned and run in consultation with Provincial Educational authorities, School Broadcast Consultation panels, local educationists, representatives of Provincial Governments and Headmasters' Associations. In May 1952 a School Broadcast Unit was set up at Karachi Station and school broadcast programmes began on 19 January 1953. The programme was recommended by a UNESCO expert who worked in Pakistan for over two years. Later, the programmes were extended to other stations.

These programmes were directed at middle classes only, i.e. 6, 7 and 8 classes. The subjects covered were History, Geography, Civics, Health and Hygiene. The broadcasts consisted of dialogues, feature programmes and plays complementary to the prescribed syllabus. There were four committees of subject teachers approved by the Director of Education, Karachi, who advised the Regional Director, Radio Pakistan Karachi on the planning of the schedules. School Broadcasts' terms were July to November, and January to March. The programmes were broadcast on four days a week for 20 minutes.

There was a demand for re-broadcast of the programme in the afternoon for the benefit of students of the afternoon-shift and this demand was accepted towards the end of the decade.

During 1965-66 a study of the methods and system of educational broadcasts in various countries like the United Kingdom, Australia, Japan, Canada and some countries of Africa and Asia was undertaken. On the basis of this study an exhaustive survey of the existing educational broadcasts of Radio Pakistan was carried out and recommendations for their reorganization were prepared. A system of registration of participating schools in each region was started in consultation with the educational authorities in order to determine

Radio Pakistan's audience and their special needs. During 1968, Karachi, Lahore, Hyderabad, Quetta and Peshawar had their own school broadcasts, six days a week from Karachi and twice or thrice a week from others. In East Pakistan all stations relayed programmes prepared and broadcast by Dhaka Station.

These programmes for the middle and high school classes were broadcast for 30 minutes daily except Friday and Sunday. At Karachi and Lahore the lessons were repeated in the evening for the benefit of students of the second shift. (In Karachi and Lahore schools normally run two shifts). These broadcasts were based on the prescribed curricula as well as enrichment programmes. The subjects covered in these programmes included history, geography, civics, home economics, health and hygiene, Urdu, English, Islamiat, economics, general science and general information.

The programmes for schools were planned in consultation and with the collaboration of local education authorities and subject expert committees and Radio Pakistan's School Broadcast Units. The General Advisory Committee of School Broadcasts (on which the Department of Education and Schools was represented) approved the programme schedule. The presentation and production of these programmes was, however, the responsibility of Radio Pakistan.

While these programmes gave teaching a new dimension they also gave students a chance to assimilate the lesson and get additional information about the particular subject beyond the syllabus. The programme was designed and presented to suit organized group listening in the school with the guidance of the teacher. The object was to supplement the efforts of the teacher. The teacher prepared the class with preliminary information in advance of the radio lesson. The broadcast was followed by a discussion and question period. Accompanying printed or cyclostyled literature was supplied in advance to all listening schools registered with the School Broadcast Unit at a station. This material consisted of term programme schedule of school broadcasts for different classes showing the subject, duration and time of broadcast; fortnightly date-chart for school notice board; and notes for teachers. At Karachi Station, Radio Clubs were formed to enhance the general knowledge of the students.

In addition to these instructional programmes Radio Pakistan also puts out enrichment programmes to provide useful information on matters of national or general interest. They provide motive to study, sharpen the capacity of the student listener to absorb information, and stimulate him/her to further pursuits of knowledge. The programmes deal with topics not included in the syllabus.

To ensure effectiveness and maximum utility of educational broadcasts feedback was sought from headmasters and subject teachers. In the light of these comments subsequent broadcasts were modified.

To provide this service to schools, Radio Pakistan had been making elaborate and costly arrangements. Its success and usefulness, however, depended entirely on the cooperation of educational authorities. They were expected to provide a good radio receiving set at each school; integrate a compulsory period in the timetable of all classes exclusively for school broadcast programmes and provide a utilization report to the School Broadcast Unit of Radio Pakistan. However, after some time it was felt that educational institutions were not availing of these programmes. This apprehension was confirmed by audience research surveys.

One study showed that the arrangements agreed upon were not being implemented and that the programmes were not used except in a few notable schools. There was no mention of these broadcasts in the school timetables. The study, however, indicated that a number of students regularly listened to these broadcasts at their homes. Some of them ascribed their success in the examination to these programmes. In view of the institutions' lack of interest it was decided in 1971 to discontinue these programmes from all stations except Karachi and Lahore. These two stations also suspended the programmes from 1 April 1975.

After a lapse of about two decades school broadcasts were taken up at Peshawar Station, under a new format known as 'Interactive Radio Instruction' (IRI). Experiments carried out in some countries of Latin America, Asia and Africa indicated that radio can play a very important role in spreading literacy. Experiments in Honduras, Dominican Republic, Bolivia, Lesotho, Papua New Guinea and Swaziland produced very encouraging results. It is found to be less expensive and yields better results than the usual educational system. Before undertaking preparation of lessons a survey was carried out in some cities and far-flung areas of NWFP. In the light of the findings a full course of 120 lessons was produced. Also, 30 lessons were prepared for the training of teachers under the title 'Calling All Teachers'. This was followed with instructional programmes for the students. In September 1993 Radio Pakistan Peshawar started broadcasting Level-1 English lessons for class-III. Level-2 programme was started in

1994 for class IV, while Level-3 programme for class V was started in September 1995. The outcome was so encouraging that it was proposed to enlarge the scope of the programme to include the teaching of science, mathematics and Urdu.

Programmes for College and University Students

Programmes for college and university students were broadcast from Lahore, Peshawar and Dhaka in the form of composite magazine programmes every week from the time of AIR. These programmes were introduced from different stations of Radio Pakistan as a supplementary teaching through the medium of broadcasting. The aim was to keep students gainfully occupied and provide them an alternative channel for self-expression. A notable feature of these programmes is the series of extension lectures by eminent educationists put over immediately before examinations. The purpose of this exercise is to help students benefit from the knowledge and experience of well-known experts and teachers on subjects from humanities and social sciences to physical sciences.

'Students Calling' the first half an hour weekly programme began on 22 December 1950 from Karachi Station. Each station broadcasts University/College Magazine or youth forum programme intended to discuss youth problems, to guide them in their studies and careers, to cultivate in them a sense of patriotism and to prepare them for future responsibilities. During 1966, the frequency of these programmes was increased from monthly to weekly and active participation of students in radio programmes was emphasized. For three decades after that Karachi Station has been offering this programme daily for 30 or 45 minutes. Other Stations broadcast once or twice a week.

In addition to these regular programmes different stations of Radio Pakistan organize 'Students Weeks' spread over seven to ten days every year, comprising *tilawat* and *naat* competitions, Urdu and English debates, quiz programme, music competition, plays and skits, and the most popular item namely 'Adabi Adalat'—mock literary court. These weeks not only encourage the younger generation to demonstrate their talent but provides a channel for self expression, and also helps Radio Pakistan discover new talent for general programmes. Students' week was not organized during the first two decades after incorporation, the last week having been held in 1971. It was after a gap of about 22 years that Karachi Station organized *Jashn-i-Tulaba* from 21 to 30 August 1993, and it took another seven years for the next festival to be held in September 2000.

Literacy and Mass Education Programmes

A campaign for mass literacy was launched during 1987. Radio Pakistan began literacy programmes from all its Stations. The programmes were the result of innovative planning and bold execution by a team of dedicated producers. It was designed and managed by PBC from its own resources. The Audience Research Wing of PBC conducted its evaluation. The programme strategy had earlier been developed in a pilot project undertaken by Radio Pakistan in a village Phalkara in Hyderabad Division of Sindh.

According to the strategy each programme has to be area-specific and participatory in approach. It should focus on one problem and concentrate in a limited geographical area. The programmes have to remain coordinated with on-going work in the field, and involve local community groups and leaders of opinion. The programmes have to be broadcast at a regular frequency to ensure continuity and exposure over a sufficiently long period of time. The programmes have to utilize new ideas and new formats. Besides folk forms (compering) and conventional formats (talks, interviews etc.), the use of drama format is encouraged and special efforts have to be made to involve urban and rural housewives in the programme.

Earlier Radio Pakistan had launched a programme entitled 'Alif Ujala' from all stations. The programme was designed to enlist and motivate teachers and students and present a report on the work done. This programme was refined and spurred the revival of hundreds of literacy centres all over the country that were functioning inadequately. Many people were motivated to start learning at the formal literacy centres.

This concept was reviewed by the broadcasting organizations in South Asia as well as at the Asia-Pacific Institute for Broadcasting Development at Kuala Lumpur, Malaysia in December 1984. The 'Alif Ujala' initiative and its bold execution indicated that Radio, apart from informing, educating and entertaining its listeners, is also capable of playing a vital role in social uplift. It has the power to reach remote rural areas faster than any other medium. Its low cost and repeated contacts, compared with the cost of contacts through conventional methods, make the radio programmes doubly attractive for the promotion of literacy and adult education. Most stations of Radio Pakistan broadcast 10 to 15 minute programmes on functional literacy.

Teachers' Training

The syllabus for primary classes was changed in the seventies. For the orientation of primary teachers in the new syllabus a 15 minute programme in Urdu was broadcast from all Stations twice a week during 1976-77 in collaboration with the Allama Iqbal Open University. The course was aimed at acquainting the primary teachers with the philosophy underlying the introduction of the new syllabus in the country. There were reports that this experiment has been successful in achieving its objective.

During the same year PBC stations in the Punjab viz. Lahore, Rawalpindi and Multan broadcast a 20 minute programme daily for students of primary schools for class I to class III according to the new syllabus.

The Mumtaz Hasan Broadcasting Committee recommended that regular school broadcasts be started from all stations of Radio Pakistan and listed the staff needed for each unit. They wanted the standard of the university students' programmes to be raised. The Committees of University professors and teachers involved in programme planning were thought to be ineffective in designing successful programmes. They recommended that listening clubs be formed in every college and contact be maintained with Radio Pakistan.

WOMEN'S PROGRAMMES

Women's role in nation-building cannot be over-emphasized. Their importance to a nation's progress is so vital, yet they form one of the most neglected sections of the population. In view of the exceptionally low literacy rates among women especially in the villages, urgent and intensive efforts were needed to begin education programmes aimed at the women-folk.

There was need to enable women to discuss their particular problems and to prepare them to make their full contribution to the advancement of society. All stations of PBC have special weekly programmes for women, generally prepared and presented by women themselves under the supervision of Radio Pakistan staff. The idea underlying these programmes is to help women become better sisters, mothers, wives and better Pakistanis, and to enable them to realize their rights and obligations as members of a society.

While some women disagreed with the idea of a separate 'compartment' for women on the air, many were pleased to have a radio programme specially designed for them, at a convenient time of the day. These programmes consist of the usual type of items in radio programmes i.e. talks, discussions, recitations, songs, features and plays, but mainly they were self-help, personal development programmes. Topics dealt with include housekeeping tips, first aid, beauty hints, child care, special news bulletins covering activities of women around the world, cooking, health and nutrition, careers for women, problems of working women, and many other similar topics of interest to women.

Programmes covering social and cultural activities of women in the country are also broadcast. Special events like women's conferences and rallies of Women National Guards and Girl Guides have either been relayed or their proceedings reported in the form of eyewitness accounts. *Musha'aras* featuring leading women poets are a popular item of entertainment. Women's institutions have always been closely associated with the planning and presentation of these programmes.

The United Nations declared 1975 as International Women's Year. Radio Pakistan started a 30-minute daily programme for women from 1 April to 31 December 1975. In these programmes the role of women in a progressive society and a developing country was highlighted and the problems of women-folk in general were discussed. International women's weeks are observed in a befitting manner from all stations.

From 1 January 1976 PBC reverted to the normal pattern of weekly programmes for women. Programmes are broadcast for educated working women and urban and rural housewives. The latter are included in the regional language programmes for rural listeners. During the last two decades or so the programme-fare for women has improved with many talented and educated women volunteering to participate.

In addition to the weekly programmes, a daily one-hour programme in the national hook-up under the title 'Hamari Dunya' has been started since 16 April 1999. In 2000, Radio Pakistan produced twenty plays on legal rights of women in collaboration with the Pakistan Women Lawyers Association. These were broadcast in the national network programme 'Hamari Dunya' every Wednesday. The themes of the plays revolved around legal issues concerning women in Pakistan, such as family laws, family relationships, marriage especially the standard form of *nikahnama* (matrimonial document), divorce, custody of children, inheritance, identity cards, workers rights, etc.

The Mumtaz Hasan Broadcasting Committee pointed out that women's programmes have undergone very little change since their inception. The emphasis has always been on home-making topics. They seemed to be directed at an urban female audience of married women. The Committee criticized the programming saying that in spite of women's emancipation and working life status, the programmes were mainly concerned with things like 'how to make a chutney'. The Committee suggested that all regional stations should start independent women's programmes in the language of their respective regions.

CHILDREN'S PROGRAMMES

Before Partition all the radio stations broadcast weekly audience participated programmes for children. These consisted mainly of fairy tales, humorous skits and songs. Soon after Independence the programme frequency was increased to thrice a week. When Karachi went on air in 1948 it also started with three children's programmes a week. The purpose in increasing the frequency of these programmes was to cater to different age groups of children. These programmes were meant for children up to 12 years of age. From January 1951, children were divided into two age groups–age 5 to 8 and 9 to 12. Different types of programmes were produced for each of these groups. However, a short while later stations reverted to producing one programme for children up to the age of 15 years.

In spite of the lack of talent and experts for children's programming, these programmes were kept on the air. New writers were cultivated. The emphasis was in providing information and showing young listeners the fundamental change that had been brought about in their lives through their country's independence.

Main features of children's programmes have been Islamic history, Pakistan history and geography, issues of importance to youth, their education and development. Competitions, quizzes and children's music and stories are popular features.

Children's programmes are now broadcast once a week in the national language and once a week in the regional languages by all regional stations. The duration ranges from 30 minutes to an hour. The programmes are presented by stock characters such as *Bhaijan, Apajan, Baji* or *Mamoonjan*, etc. In the first few years after Independence, Lahore Station broadcast a programme 'Al-Mustaid' for Scouts and Girl Guides. Some stations broadcast bed-time stories usually with a moral theme. In general, Radio Pakistan's children's programmes groom young listeners to be good Muslims and good Pakistanis.

The Mumtaz Hasan Broadcasting Committee was of the view that children's programmes were 'affected and have a quality of sameness'. Ever since the inception of broadcasting in the subcontinent, these programmes have remained more or less unchanged in technique and presentation. The Committee suggested that the programmes should be divided on the basis of age groups and should be planned in consultation with child psychologists and educationists.

Forces Programmes

On 1 December 1950 the first programme dedicated specifically for Pakistan's Armed Forces was inaugurated. The Commander-in-Chief of the Pakistan Army at Rawalpindi Station performed the inauguration. Lahore Station included special items for the troops in its rural programmes. Karachi Station used to broadcast a bi-weekly programme that was relayed by all Stations. From 15 June 1965 the Forces programme was aired on a regular daily basis from all stations in the two Wings of the country. The Commanders-in-Chief of the Army, the Navy and the Air Force recorded inaugural messages on the occasion. The programme includes talks, interviews with ex-servicemen and relatives of soldiers, national songs, band music, news, features and skits. It is presented by stock characters who have a background in military service and pleasant on-air personality.

During the 1965 emergency, the scope of the programme was expanded. Now it includes messages of Defence personnel from various units supplied by GHQ/ISPR, talks on the history and achievements of different regiments/units, eyewitness accounts of social, cultural and other events, special news bulletins, inspiring features, patriotic, martial and film songs and documentaries stressing the martial traditions and heroic role of the Forces in guarding the national frontiers. Stations receive encouraging feedback from soldiers about these programmes. Liaison is maintained with ISPR to get the reaction and comments of the Armed Forces personnel. Most stations schedule this 30-minute programme in the evening.

The Mumtaz Hasan Broadcasting Committee said that the Forces programme lacked the 'feel of the barrack room' primarily because the programme is planned and produced by Radio Pakistan without the guidance of the Armed Forces. The Committee recommended that the Inter-Services Public Relations Directorate should establish contact with Radio Pakistan at a central level and feed the organization with live and topical material regularly.

PROGRAMMES FOR LABOUR

Programmes for industrial workers are broadcast from different stations to discuss labour problems, to inform and educate them on their rights and privileges under labour laws and under the Constitution. The aim is to make them aware of labour reforms and inculcate dignity of labour. Legal advice is also given. Besides entertainment they are provided motivation to work hard for the greater good of the country. The objectives of the programme also include: labour-management relations, encouraging healthy trade union activities and creating conditions conducive to accelerate the pace of industrial growth and boosting the national economy.

Most stations air 10 to 30 minute programmes for workers on a daily or weekly basis.

STATION DIRECTORS CONFERENCE

Fielden was very keen on periodical meetings of the Station Directors. These meetings were held every quarter and if possible at a newly established station. After some time these conferences were scheduled after six months and still later became annual events. At these conferences the annual performance of each station is reviewed, current programmes are discussed and plans formulated for the future.

Problems come up for discussion on administration, programme planning and production, availability of talent, exchange of programme material and artists and other topics. The conferences give Station Directors the opportunity to interact and build cooperation. These conferences are presided over by the Director-General. The Minister for Information and Broadcasting, the Information Secretary and other officials of the Government attend by invitation. The minutes of these conferences are circulated to all Stations for action as laid down therein.

After Independence, the first Station Directors Conference was held in Lahore in January, 1948; the second at Karachi in December, 1948; the third in December, 1949; the fourth in September, 1950; the fifth in July, 1951; the sixth in April, 1952 and the seventh in January, 1953, all at Karachi.

The first five Conferences were addressed by the Minister for Interior and Information and Broadcasting, Khwaja Shahabuddin, and the last two by Dr I.H. Qureshi, Minister for Information and Broadcasting. The 15th Directors Conference was held, for the first time, in East Pakistan (Dhaka) in March 1966. Sometimes conferences are held on a regional basis. The Station Directors Conference held in October-November 1995 was organized after a lapse of more than four years. The previous one was held in September 1991. The 1995 Conference dealt with such matters as revival of the old tradition of drama festivals, student weeks and *mehfil-i-musiqi*; audience-participated programmes, greater coverage of rural areas, more attention to music production at stations, FM channel programmes, measures to increase PBC revenues, revival of PBC awards and presentation section, inter-station competitions, review of the list of anniversaries and festivals etc.

The Directors Conference held in March 1999 set up six working groups which thoroughly discussed such items as the existing system of newscasts, plays, music, humorous programmes, women's programmes, morning national hook-up programme 'Subh-i-Pakistan' and FM 101. Some of the recommendations of the conference were:

1. The pattern of the 8.00 p.m. news bulletin be changed. It should be *Khabarnama* of half an hour including national and international news, finance and commerce, sports and special weather reports, voices of Radio representatives etc.

2. The 50-minutes morning programme 'Subh-i-Pakistan' be reintroduced with definite changes from 23 March 1999 with a desirable balance between information and entertainment.
3. One-hour programme for women only under the title 'Hamari Dunya' be broadcast from the seven bigger stations at 10 a.m. from 1 April 1999. Islamabad Station to give lead.
4. Radio drama be revived. All big stations should broadcast serial plays on weekly basis. These must have some purpose/message. A drama festival be organized in September 1999.
5. Keeping its old traditions, humorous programmes should be introduced. Karachi, Lahore and Peshawar should start weekly programmes.
6. A live *musha'ara* should be broadcast every month from Islamabad and the four provincial capital stations.
7. More rural-oriented programmes be broadcast particularly from regional stations, in which participation of listeners is to be given special attention.
8. Radio documentary which was once a popular programme be reintroduced.
9. Radio Pakistan to broadcast special programmes on the first anniversary of atomic explosion. The main point to be emphasized in these programmes is the deterrent capability of Pakistan to meet any external aggression.
10. Special programmes to be broadcast on the bicentennial anniversary of Tipu Sultan Shaheed.
11. FM 101 transmission to be expanded. Its coverage be increased from Karachi to Hyderabad and from Islamabad to Peshawar.
12. *Jashn-i-Baharan* Radio Festival be revived, under which music concerts to be held at different stations culminating at Islamabad in which participation to be ensured of artists who attained prominence through Radio.

Similar professional meetings are also held in other wings of the Corporation e.g. Engineering, Finance and Sales. A three-day Station Engineers Conference was held at the end of November 1949. It was the first of its kind to be held in the Indo-Pakistan subcontinent.

ADVISORY COMMITTEES

Each station has an advisory committee to advise on programme matters. There is also a Central Advisory Committee at the

Headquarters. These committees are appointed to advise the PBC on all matters relating to the quality and acceptability of its programmes and convey public reaction. This was a way to associate professionals with the planning of programmes so as to make Radio Pakistan more responsive to public opinion.

These Committees have a fixed tenure of two years but are requested to continue till the appointment of the next committee. Their membership is honorary. The Committees are advisory in nature and do not exercise administrative or executive authority with regional stations. However, the advice of the Committees on matters of vital concern to the public in the field of broadcasting is almost always accepted and every possible effort is made to facilitate their recommendations.

An attempt is made to obtain as wide a representation of cultural, educational and social interests on these committees as possible. The members come from the academic world, business, law and parliament, religion and special groups like labour, rural population, women and minorities. The Central Advisory Committee meeting is chaired by the Minister for Information and Broadcasting while the regional committees are normally presided over by the Director General. They meet periodically, at least once a year. A complete report of the activities of the regional station is placed before the committee. The other subjects on the agenda may include; matters of special interest to the public such as proportion of broadcasts in regional and national languages, relationship of programmes to audience responses, contents of religious programmes, programmes for special audiences like students, women, children and rural listeners, organization of community listening and so on.

The Advisory Committee can appoint where necessary subcommittees to investigate particular problems. The members speak freely on the performance of different sections of PBC and express their views regarding the possible improvements that can be effected. For example during the two sessions held during the year 1974-75, some of the recommendations made by the Central Advisory Committee related to: preparation of programmes to observe centenaries of the Quaid-i-Azam, Allama Iqbal and Hazrat Amir Khusrau; starting of documentary programmes on Pakistani cities, improving the language of commercials; broadcast of *musha'aras,* programmes on family planning, minorities and women; weekly programmes to explain Acts of Parliament; and music on FM channel.

There are special subject and expert committees at most stations, who advise on matters specific to their field, viz. religious broadcasts, music, educational and farm broadcasts, talks, plays and so on. These committees are composed of experts as well as well-informed listeners interested in a particular category of programmes. They are intended to advise the Station Director, who acts as convenor of each committee. These committees have been far more active than the advisory committees and have brought attention to listeners' demands in programmes.

The tenure of the last Central Advisory Committee expired on 31 December 1983 and was extended by one year. Since then there has been no Committee at the Headquarters or the Regional Stations.

8

Overseas Broadcasts

External Services

In British India foreign broadcasting began during the Second World War with a sense of urgency as the initial success of the Axis countries posed a growing threat to the British Empire. A month after the declaration of war on 3 September 1939 a news bulletin in Pushto was started from Delhi. It was aimed at Pushto-speaking Indians as well as listeners across the border in Afghanistan in an effort to neutralize Nazi propaganda. A bulletin in Afghan-Persian (Dari) was started from Delhi on 4 December.

Japan's entry into the War in 1941 posed a new threat to Britain's Far Eastern territories. The Government of undivided India decided to install a number of high power short wave transmitters on high priority basis. The first of the five 100 KW SW transmitters went into operation on 1 May 1944. By 1945 AIR's external programmes had grown to 74 daily in 22 languages. When the War ended the number of daily broadcasts was reduced from 74 to 31 by early 1947.

At the time of Partition the entire external broadcasting complex was located in India and the Indians refused to share it with Pakistan. Pakistan's broadcasting authorities had to make urgent efforts to procure transmitters and other essential equipment and recruit and train staff in order to function and fulfil its obligations as determined by the Government and later specified in the Pakistan Broadcasting Corporation Act of 1973.

The need for projecting Pakistan to the outside world was keenly felt at the birth of the new state. Pakistan was still an unknown country with no voice to the world. Indian leaders had predicted that Pakistan was not an economically viable entity and was bound to revert to being a part of India. There were those who were waiting to see how the young nation would survive unprecedented odds and they had to be shown that Pakistan would overcome its initial trials and survive.

There was another reason for the urgency to begin broadcasting operations. Pakistan had risen with a new way of life and had a message of peace and cooperation for a war-weary world. There was, therefore, a need for the voice of Pakistan to reach out to other nations.

Under the PBC Act of 1973, the Corporation is required to broadcast programmes in the External Services according to the directives of the Government with regard to range, language and timings. PBC has, therefore, provided the principles and policies of Pakistan to foster better relations with other nations. The broad objectives of the External Services are:

i) To furnish authentic information about Pakistan's rich historical past and its present struggle to transform itself into a modern, democratic and progressive Islamic society.
ii) To strengthen brotherly ties with the Muslim countries and forge close relationships with other Third World countries, members of NAM and SAARC in particular and all other countries in general.
iii) To support struggle of self-determination of subjugated people, wherever they may be and the efforts to build more equitable international order.
iv) To present Pakistan's social/cultural heritage and foster friendship with other countries by stressing common ties and affinity of interest and aspirations.
v) To portray Pakistan's progress and to focus attention on the economic problems of the Third World and the need for more free trade relations among nations.
vi) To counter any biased or hostile propaganda against Pakistan by circulating objective facts.
vii) To win more and more friends for world peace and all just causes.

Due to innumerable problems and especially the lack of adequate resources and facilities, Pakistan could not start its External Services for two years following Independence. A 50 kilowatt short wave transmitter was installed at Landhi in Karachi on 14 August 1949 and from the same date four Services were started. These were Afghan-Persian, Arabic, Iranian and Burmese (later Myanmar) Services. Afghan-Persian, Iranian and Burmese Services were of 45 minutes each, while the Arabic Service for technical reasons was broadcast in two chunks of 45 minutes each. The first chunk of the Arabic Service was directed to Arab countries east of the Suez Canal, i.e. Palestine, Lebanon, Syria, Iraq, Jordan, Saudi Arabia, the Yemen and Gulf countries; while the other chunk was meant for countries to the west of the Suez, i.e. Egypt, the Sudan, Libya, Eritrea, Tunisia, Morocco and other North African countries.

All the four Services were broadcast on the same frequency (11885 KCS in the 25 metre band) on the same transmitter. The Afghan-Persian Service was, however, carried by the medium wave transmitter of Peshawar Station as well, as a regular programme since 1942 for listeners in Afghanistan, North West Frontier Province, Balochistan and Tribal Areas on both sides of the Durand Line. It may be of interest to know that the Afghan-Persian Service was the last item broadcast from Peshawar Station of AIR at the time of Independence. During that period the Station's main purpose was war propaganda and countering of foreign influence in the area. After Independence it aimed at presenting a true picture of the situation in Pakistan and attempted to foster neighbourly ties with Afghanistan. Also to enlighten the Tribesmen about the rest of the world, help them eradicate social and moral evils that were hampering their progress, and play a role in the building up of a new Islamic society.

In this Service music was performed by male singers only due to strict tribal orthodox views on female voices being heard in public and on radio. Despite the anti-Pakistan propaganda consistently carried by Kabul Radio from the time of Partition, PBC's External Service never resorted to similar attacks. Kabul Radio propaganda was aimed at creating anti-Pakistan feelings amongst the Tribesmen, urging them to resist change and progress and retain their primitive lifestyle, as they did under the British. To resist the spread of education, building of roads, hospitals, factories and such other amenities of life. Their loyalty to Pakistan was challenged.

From January 1950 however, it was decided to make a dignified rebuttal of the baseless allegations and vituperative propaganda emanating from Kabul Radio. Radio Pakistan's broadcasts were well-received by the Afghan people and Tribesmen and it became so effective that listening to Radio Pakistan was banned throughout Afghanistan. The Service proved its worth and utility during the Russian aggression against Afghanistan.

The following five Services were started on a purely experimental basis on 18 January 1952:

1. South East Asian Service, broadcast in the early hours of morning for 45 minutes.
2. South Asian Service, broadcast at midday for one hour and 15 minutes.
3. Indonesian Service, broadcast late in the afternoon for 45 minutes.
4-5. Two midnight transmissions directed to Turkey and the United Kingdom for 45 minutes each.

Commercial records were played in these broadcasts. By 18 September 1952 the word experimental was dropped and a news commentary or a news talk of five to ten minutes was introduced, to present Pakistan's viewpoint on events inside the country. The East and South East African Service, which was also being run on experimental basis was made a full-fledged service from 10 March 1952. This Service enjoyed immense popularity with African nationals of Indo-Pakistan origin and Gujrati-speaking peoples.

With improvement in the availability of resources and setting up of short wave transmitters new services were added. With the commissioning of two 250 KW short wave transmitters at Islamabad in 1972-73, it became possible to plan the External Services on a more rational basis. Some of the services were only of 15 to 20 minutes duration.

The External Services is headed by a Controller and a Deputy Controller. External Services up t 14 August 1997 were being broadcast for 20 hours a day in fifteen languages. These were Arabic, Dari, French, Gujarati, Hazargi, Hindi, Indonesian, Iranian, Mitali (Bangla), Myanmar, Swahili, Sylheti, Tamil, Turki and Turkish Services. In addition, a 45 minute programme in Pushto entitled 'Hindara' is broadcast daily from Peshawar Station on medium wave for Pushto-speaking listeners across the border.

Nepali Service was started in October 1970. It originated from Dhaka Station of Radio Pakistan. When Pakistan was dismembered, this service continued to be broadcast from Radio Bangladesh, Dhaka. The Sylheti Service was started on 1 October 1971. Besides Sylheti-speaking population in East Pakistan it also catered to Sylheti workers in the United Kingdom. It was later made part of the Mitali Service. The Mitali Service (Bangla) was started with one-hour duration in the Home Service in December 1971 during the war with India to provide listeners in East Pakistan with information about the situation in West

Pakistan. Later on, it was decided to continue the Service to provide a bridge of communication with listeners in Bangladesh. It became part of the External Services on 14 April 1972.

The policy of PBC has been to promote better understanding between the peoples of Bangladesh and Pakistan. The positive aspects of the past relationship between the two former Wings of the country and Pakistan's desire for future cooperation were highlighted. On the same date, i.e. 14 April 1972 the duration of the Service was increased to two hours and further to four hours from 1 January 1974. At the beginning of 1997 it was on the air for three hours in two chunks. All other Services were of one or two hours. The Dari Service is picked up by the medium wave transmitter at Islamabad Station.

The year 1997 saw a number of changes in External Services. The Myanmar Service was stopped from 4 May 1997 while Swahili and Sylheti Services were discontinued from 14 August 1997. Two new services, Chinese and Russian, were added on 12 May 1997. These are directed towards China and the Muslim States in Central Asia. The duration and timings of almost all the services were changed with effect from 14 August 1997. The Hindi Service was of 90 minutes and Dari Service of one hour. Arabic, Irani and Mitali Services were of 45 minutes each. The duration of the remaining nine services (Chinese, French, Gujarati, Hazargi, Indonesian, Russian, Tamil, Turki and Turkish) was 30 minutes each. PBC thus broadcasted 14 services for a total duration of 9 hours and 15 minutes daily.

From January 1999 a new Assami Service was introduced in English for half an hour and the duration of Bangla, Hindi and Tamil Services was increased bringing the total duration to 11 hours. Two more Services, Nepali and Sinhali were added from 26 March 2000 each with a duration of half an hour. As such PBC now broadcasts 17 services in as many languages for a total of 12 hours.

The External Services of Radio Pakistan are directed to about seventy countries in South East Asia, Central Asia, Middle East, North Africa and Europe. Liaison is kept with Pakistani Missions abroad and their suggestions are sought for improving programmes and making the Services more purposeful for the target areas.

The programme-fare of the External Services in general consists of recitation from the Holy Quran, news bulletins, backgrounders to news, interpretative commentaries, features, special programmes on important national and international events/days, sports round-ups, talks, discussions, dialogues, interviews, plays and musical presentations.

In the beginning programme details of the External Services were published in the English journal 'Pakistan Calling' and *'Sada-i-Pakistan'* (Persian), *'Azaat-i-Pakistan'* (Arabic) and *'Pakistan Athan'* (Myanmar). The last three were discontinued. Details of programmes, timings, frequencies and meter-bands of all External Services are published in the programme bulletin 'Pakistan Calling' issued by PBC Publications from Karachi every month in the form of a pamphlet for free distribution among listeners in the target areas. The programme details are in six languages—Urdu, English, Arabic, Iranian, Dari and Turki.

WORLD SERVICE

There was criticism that the External Services while being Pakistan's ambassador of peace and goodwill, did not quite meet the needs of Pakistanis living abroad who felt a sense of detachment from their native land. This need was addressed in the one and a half hour World Service that began on 21 April 1973 which aimed to keep overseas Pakistanis informed of developments at home, and to project Pakistan's point of view to both Pakistani and non-Pakistani listeners in the target areas.

World Service, headed by a Deputy Controller, specializes in programmes that keep Pakistanis abroad aware of what is happening in the home country. These programmes cover such diverse topics as national affairs, the economy, music and lifestyle changes. It elaborates on Government incentives to attract foreign investment in the private sector, identifying areas of prospective industrial and technological development. To facilitate overseas Pakistanis travelling home, information about Customs regulations is a regular feature of the Service.

The World Service originated from Karachi with two transmitters of 50 kilowatt and with two transmitters of 250 kilowatt at Islamabad. In August, 1984 it started broadcasting from Islamabad. The World Service is broadcast in five chunks totalling 10 hours for listeners in the following regions:

Timings (UTC)	Target Areas
1. 0100 – 0200	South East Asia
2. 0500 – 0700	Gulf and Middle East

3. 0800 – 1100   Western Europe and UK
4. 1330 - 1530   Gulf and Middle East
5. 1700 – 1900   Western Europe and UK.

The World Service is broadcast in Urdu and English. All the five chunks start with recitation from the Holy Quran and the programme-fare generally consists of *Hamd*, *naat*, news, news commentaries, backgrounders, sports round-ups, talks, interviews and entertainment items. The first chunk is a relay of Islamabad Station. In addition to these, the World Service carries a special *Haj* Service for the benefit of *Hajis* in Makkah and Madina, for a period of about five months before and after the *Haj* festivities.

Live coverage of international sports events, whenever possible, are also relayed by this Service for the benefit of Pakistani listeners. The World Service broadcasts a 30-minute weekly programme 'Ham Watan' (previously under the title 'Door Baithe') in cooperation with the Overseas Pakistanis Foundation, in which problems of Pakistanis living abroad are discussed. It is one of the most popular programmes of the World Service.

Feedback about the External Services is available in the form of reception reports from Pakistan missions/embassies and foreign broadcasting organizations as well as Dxers requesting QSL cards. These comments are not regular, however, and mainly concern the technical quality of reception only. Another source of feedback is listeners' letters that relate to both programme contents and quality of reception. No audience research survey has ever been done to find out the quality and extent of reception in the target areas. There was a time when listeners' mail ran into thousands, but this has dwindled considerably.

Since External Services are broadcast on short wave only there is a possibility that reception in target areas is inaudible. The local population in the target areas has better sources of information in the form of home stations and stations located in the vicinity of their homeland. Content of a service also has an impact on listenership. Foreign news broadcast from Pakistan may be pointless, as listeners can get up-to-date news quicker from other sources. Timings of broadcasts are equally important, and should be accordin to the convenience of listeners and not when the transmitters can be spared from home or other priority requirements. Staffing and remuneration rates cast their influence on the quality and acceptability of the service.

Qualified persons (both local and foreign) are not normally available in the country. PBC often employs foreigners in Pakistan on contract basis. This stop-gap arrangement is neither satisfactory nor dependable. As a rule the head or supervisor of a service needs to be a Pakistani and the staff can be Pakistani or foreigners who have command of the language and speak in the accent and idiom of the target area, to be effective. With the remuneration offered by PBC it has not been able to attract suitable talent.

All these factors tend to compromise the quality and listener-appeal of a service. There is thus a need for increasing the signal strength, improving the contents and re-scheduling the time of broadcasts to improve the quality and listenership of World Services.

9

NEWS AND CURRENT AFFAIRS

The state of news broadcasting at the time of Independence has been described in Chapter 3. To recapitulate, there was no short wave transmitter and no centralized news system. There was severe staff shortage and too few trained persons to successfully run the news service at three locations. News channels had ceased, riots had led to a complete breakdown of news sources and essential equipment like creed machines, typewriters, wire services, and even furniture and stationery were not available. In these circumstances news was broadcast separately from the three units located at Lahore, Peshawar and Dhaka. Reviving and running an adequate news service was, therefore, the greatest need of the hour.

With these severe limitations, Radio Pakistan began a news broadcasting network called the Central News Organization. Dhaka started with four bulletins in English and three in Bengali. Lahore Unit prepared four bulletins both in English and Urdu. Peshawar Station relayed these bulletins by telephone. Lahore also edited three bulletins in Pushto that were broadcast from Peshawar. Peshawar Unit was responsible for one 10-minute bulletin in Afghan-Persian for Tribal areas and Afghanistan. All told the three Units were putting out 19 bulletins in five languages. Gradually as things started settling down more bulletins were added. A fifteen-minute Urdu bulletin was started by Dhaka in December 1947. A Kashmiri bulletin began from Lahore during Indian aggression in Kashmir.

When Karachi came on the air on 14 August 1948, it started relaying news from Lahore by telephone just like Peshawar. A year later, the initial difficulties were partially overcome with the commissioning of short wave transmitters for Radio Pakistan. The centralization of the news service began and the bulletins were broadcast from Karachi for listeners in Pakistan and also neighbouring countries. A uniform news standard and policy were adopted. The News Units were retained, though at reduced strength, for the purpose of gathering news and feeding regional stories to the Centre. Encouraged by this success the News Unit at Lahore was revived and more units created at Rawalpindi and Quetta. The Quetta Unit was, however, closed after sixteen months due to financial difficulties.

An exclusive Radio Pakistan teleprinter circuit was established over West Pakistan to link the regional news units at Lahore, Rawalpindi and Peshawar to Karachi to facilitate quick and adequate flow of news items. Other sources of news included radio's own newsgathering team covering major events, e.g. budget session of the Assembly, proceedings of Parliament and the Constituent Assembly, social and cultural functions, Central and Provincial Information Bureaus, monitoring reports prepared by CNO News Listeners, and subscribing to five world and two national news agencies.

Instead of a dearth of news material, the problem was that of sifting, selecting and condensing thousands of words received from different sources into inflexible limits of 10 or 15 minute broadcasts. In March 1954, Radio Pakistan broadcast 435 minutes of news every day through 42 bulletins—national, regional and local – in English, Urdu, Bengali, Pushto, Sindhi, Balochi, Gujarati, Kashmiri, Balti, Shina, Pahari, Afghan-Persian, Iranian, Arabic and Myanmar for listeners extending from North Africa to the Far East. At that time 26 broadcasts of a total duration of five hours and thirty minutes were monitored every day from such stations as BBC, AIR, VOA, Voice of Indonesia, Radio Australia, Radio Moscow, Radio Damascus, Radio Japan, Radio Ankara, Radio Tehran and Radio Peking (now Beijing).

After the 1965 war between India and Pakistan some radical changes were made in the working of the Central News Organization, as in other departments of Radio Pakistan. In the light of the new broadcasting policy announced after the War on 23 October 1965, CNO reinforced its news gathering machinery and accelerated the pace of its news services. The number of News Units rose from six in 1958 to twelve in 1968. During the same period the number of news bulletins went up from 50 in 16 languages to 81 in 21 languages. Although it is a medium of education and information, Radio Pakistan is susceptible to pressure of public opinion, yet aims to galvanize people for the achievement of national objectives. Keeping this cardinal principle in view Radio Pakistan adopted a comprehensive programme to improve the contents and presentation of its newscasts.

As a vast majority of Pakistanis are not even acquainted with English it was decided to reduce the number of news bulletins in that language. The English bulletins broadcast in the national hook-up at 1.10 p.m., 5.10 p.m. and 10.05 p.m. were discontinued from 15 November 1965. In addition, the local news bulletins from all Stations and the English news commentary in the national hook-up at 10.20 p.m. ceased. The English bulletins were replaced by bulletins in Bengali and Urdu. The duration of English newscasts dropped from 50 minutes to 25 minutes. West Pakistan Stations started relaying Bengali bulletins and East Pakistan Stations relayed bulletins in Urdu. Two Punjabi news bulletins were introduced from Lahore, one for East Punjab and the other for rural areas of West Pakistan. Special news bulletins were introduced in the Armed Forces programme of various stations.

News bulletins in the national and regional languages were always translated from English bulletins. This made the local news reports sound stilted and it lost much of the effect of the original news bulletins. It was, therefore, decided to draft news bulletins directly in Urdu and Bengali for broadcast from Karachi. The Hyderabad Unit also started writing Sindhi bulletins directly in Sindhi. Other regional units were expected to follow the example. Unfortunately this experiment did not achieve its desired result and was given up a little later mainly because the majority of Editors could not adjust to the change.

The most important national or international event of the day was the topic of the daily news commentary. The national topics related to political, social and economic developments. Pakistan's foreign policy moves were also interpreted.

After its incorporation PBC took two important steps to effect improvement in the working of the Central News Organization. One, the staff position was improved and secondly the transport and telephone facilities were substantially increased resulting in easier mobility for the news staff. At the time of conversion CNO handled 71 bulletins. In 1977, the bulletins began with a dedication to the Almighty Allah and the greeting of *Assalamo-Alaikum* as an indicator of Islamic culture.

On 1 November 1986 a bold change in the pattern and strategy of newscasting was made on an experimental basis. As a result news was brodcast every hour on the hour from 6 a.m. till midnight. The duration of the bulletins varied from 3-15 minutes. Simultaneously, the number, timings and duration of regional and local news bulletins also underwent a change. Four bulletins each were put on the air in Punjabi, Sindhi, Pushto and Balochi—the first and the last of seven minutes and the remaining two day bulletins of four minutes each. One bulletin each in Brahvi, Hazargi, Balti, Shina and Arabic was broadcast from Quetta, Skardu, Gilgit and Islamabad. Thirty-nine local bulletins in Urdu and local languages/dialects were broadcast from all regional stations.

The local bulletin from Islamabad is in English. About sixty minutes were devoted to the eight current affairs programmes in the national hook-up. These included three news commentaries–two in Urdu and one in English, editorials, backgrounder, commercial review, resume of Parliament's proceedings and a programme 'Gird-o-Pesh' which gave detailed analysis of the main items of the day.

In all 108 news bulletins were broadcast daily in national, regional and foreign languages for a total of about 10 hours and 45 minutes. These included 19 national bulletins, 25 regional, 41 local, two FM, 16 External Services and three slow speed bulletins. In addition, World Service relayed 10 bulletins from Islamabad. The slow speed bulletins were for overseas listeners and Pakistan Missions in South East Asia, Western Europe and East Africa/Middle East. Of the 19 national bulletins six were in English and the rest in Urdu including the commercial bulletin started in August 1996. The duration of these bulletins varied from 2 minutes to 12 minutes. The duration of the 89 news bulletins in the Home Service was eight hours and 13 minutes. The 16 External Services bulletins consumed an air time of one hour 42 minutes per day.

The daily duration of all the Home and External Services bulletins and current affairs programmes came to 11 hours and 45 minutes. It did not include the 10 national bulletins relayed by the World Service. In addition to the slow speed bulletins, a special daily news summary was sent to Pakistan Missions in New York, Washington and Ottawa for the information of the Pakistani community in the USA and Canada. Special bulletins are arranged during emergencies and important events like floods, summit conferences, elections etc.

Some changes in the national and regional news bulletins were introduced with effect from 14 August 1997. The duration of a number of Urdu bulletins was reduced to make them headline news of about two minutes. The headline news includes only hard news of national and international significance in a nutshell of two or three sentences.

Some Urdu and English bulletins were dropped to give room to regional news. Midday Punjabi, Sindhi and Pushto bulletins were replaced by 4-minute news items in these languages and broadcast in the national hook-up at 1200 noon, 1300 hrs and 1400 hrs respectively. The duration of the 2000 hrs Urdu bulletin and 'Gird-o-Pesh' was restored to 15 minutes. A weekly discussion programme was added on Fridays. The effect of these changes was that on 14 August 1997 the number of national, regional, local, World and External Services bulletins was 104 for a total duration of 606 minutes.

The main 8 p.m. bulletin was given a new format to incorporate new technology. On 23 March 1999 this bulletin was turned into a *Khabarnama* of 30 minutes giving national and international news, prominence to news from far-flung areas of the four Provinces, voice inserts of PBC representatives, telephonic interviews, opinion of people on important issues, as well as views of an expert on pressing issues, commercial news, sports news and a detailed weather report. On 1 April 1999 a 4-minute Balochi bulletin was added in the national hook-up at 1604 hours. The three general overseas slow speed (GOS) bulletins were stopped on 1 August 1999 as seven national Urdu/English news bulletins had been put on the Internet.

The news policy of PBC is to present national and international news timely, accurately and in as objective and impartial a manner as possible. Sufficient time is devoted to foreign news items and it is ensured that breaking news is speedily reported. On the home front, development news takes a major portion of the total time. Most of this news comes from various Government announcements or from the statements made by important Government functionaries or through institutional sources. Radio Pakistan gives due cognizance to this type of news because in Pakistan, particularly in villages, Radio is often the only source of information.

Among the newreaders who have been popular are: Shakil Ahmad, Anwar Behzad, Wirasat Mirza, Abdus Salam and Shamim Ejaz, for Urdu news, while among the English newsreaders are Jahanara Sayeed, Eric Warner, Rizwan Wasti, Aslam Azhar, Edward Carrapiett, Anis Mirza, Anita Ghulamali, Shaista Zaid and Riaz Ahmad Khan.

Table 1: Statistical Position of the News Bulletins

No.	Types of Bulletin	As on 13.08.1997 No. of Bulletins	As on 13.08.1997 Duration (minutes)	As on 30.06.2000 No. of Bulletins	As on 30.06.2000 Duration (minutes)
1.	National	19	114	22*	137
2.	Regional	25	169	17	142
3.	FM	2	4	7	21
4.	External	16	102	20	100
5.	GOS	3	45	-*	-
6.	Local	41	201	42	206
7.	Commercial	1	5	-**	-
8.	Arabic	1	10	-†	-
9.	News Summary	1	5	-*	-
	Total	109	655	108	606

* = Of these national bulletins seven are broadcast on Internet also. GOS bulletins and News Summary dropped for this reason.
** = Included in Khabarnama.
† = Included in External Service (Arabic).

CURRENT AFFAIRS

The current affairs section was created after incorporation and was run by a Controller. The section was responsible for daily news commentaries, current affairs discussion programmes on political, social, cultural, educational, sports and economic topics. The news commentaries are written by eminent journalists, experts and PBC staff and are translated into regional languages for broadcast. Discussion programmes (normally of 15 to 30 minutes) are arranged as and when required, in which significant national issues and international developments are discussed by prominent journalists and subject experts. During the sessions of the two Houses of Parliament a resume of the proceedings is given every evening.

The purpose of the current affairs programmes is to provide in-depth treatment and analysis of the main news items. The sphere of activity of the Current Affairs Section of the Central News Organization was very limited being confined to the broadcast of two 5-minute commentaries, a review of newspaper editorials, rehash of items for 'Gird-o-Pesh', a weekly discussion programme and resume of Parliamentary proceedings. Prior to incorporation, one or two producers did this work. In actual practice the responsibility of current affairs remains with the stations and in particular with Islamabad Station. For several years after its creation the Current Affairs Section of CNO was headed and run by programme personnel.

EXTERNAL SERVICES AND CURRENT AFFAIRS WING

To remove this anomaly and to streamline functioning, a Current Affairs Unit was created at the Headquarters to coordinate the work of CNO and the Stations. Also a separate External Services and Current Affairs Wing was created at the Headquarters on 18 August 1999 to combine the existing External Services Unit and the newly created Current Affairs Unit under a Consultant with the status of a Director. Simultaneously three cells were created at Islamabad, Karachi and Lahore Stations, each headed by a Deputy Controller and Programme Managers cells at other stations.

This Wing is generally responsible for all programme activities relating to opinion-building on major current issues of national and international politics, foreign policy projections to advance national interests and promotion of Government policies, initiatives and performances. The new set-up encompasses all current affairs activities both at the Headquarters and all major stations. At present four commentaries (three in Urdu and one in English) are broadcast daily. The current affairs sections of regional stations continue to broadcast their programmes locally or put on the national/provincial hook-up as the case may be besides feeding the same to the Unit at the Headquarters.

NEWS AND CURRENT AFFAIRS CHANNEL

To deliver breaking news from around the world, PBC decided to launch a News and Current Affairs Channel, which was inaugurated by the Chief Executive, General Pervez Musharraf on 18 April 2001 at the National Broadcasting House, Islamabad. This Channel operates in two separate chunks from 7 a.m. to 9 a.m. and from 6 p.m. to 11 p.m.

Programmes of the Channel, besides the usual news bulletins, include voice reports of correspondents, actualities, on the spot interviews, investigative reports, discussions, documentaries, features, newsreels, developments in science and technology specially in IT, book reviews, business reports, commentaries, sports analysis and weather forecasts.

PBC Lahore-II, Bahawalpur, Hyderabad-II, Quetta-II, Khuzdar, Faisalabad, Dera Ismail Khan, Zhob, Loralai and Turbat also relay the channel in addition to the new 100 KWMW transmitters at Rawalpindi and Karachi.

Monitoring

The Monitoring Department of Radio Pakistan began functioning as a unit of the Department of External Services and Monitoring on its creation in August 1949. The Monitoring Unit was manned by four optees from the Simla Monitoring Office of AIR. These four hands produced a daily report comprising the Urdu, Arabic and Persian broadcasts from Delhi and Moscow. A year later four more Monitors joined bringing the total to eight, and Kabul and Tashkent were added to the monitoring schedule.

The Monitoring Department performs a dual function. It serves as a news collecting agency and an important organ of counter-propaganda machinery, as well as an intelligence source to the Government for information on various broadcasting stations of the world. The Department proved useful from the very beginning particularly regarding Kabul's propaganda for a separatist Pakhtoonistan. It became a full-fledged Department with staff and funds allocated. In the interim an Emergency Monitoring Centre was set up in a barrack at Ack Ack School, Clifton, Karachi. Installation of technical equipment was completed in 1952 and the entire monitoring staff was shifted to the new premises by March 1953. On 12 October 1953, the Monitoring Department was separated from the External Services and became an independent subordinate office of Radio Pakistan under a Director of Monitoring. Within four years it grew into a full-fledged Department run by professional staff.

At the beginning of 1954, the Department had on its daily monitoring schedule about 55 broadcasts from 13 foreign stations in eight different languages. The Stations covered were Srinagar, Kabul, AIR, Baghdad, BBC, Cairo, Indonesia, Moscow, Beijing, Rome, Tehran and Voice of America and the languages were Arabic, Bengali, Dogri, English, Kashmiri, Persian, Pushto and Urdu. The Department produced three documents at that time. The first was the 'Daily Report of Foreign Radio Broadcasts' running into 30 to 35 foolscap pages. The second was a summary of important items that was supplied to some of the Ministries. It preceded the Daily Report and was intended to provide the Ministries concerned with important material likely to demand immediate attention and action. The third report was a weekly analysis that was considered to be the most important of the three. It involved observations on the trends of foreign stations broadcast during the week. It gave a resume of the week's important developments and a note on the tone and contents of the broadcasts.

Before incorporation there were small Monitoring Units at Lahore, Rawalpindi and Quetta in addition to the Directorate of Monitoring at Karachi. The Lahore Unit monitored Urdu broadcasts of AIR; Rawalpindi Unit was responsible for Urdu, Kashmiri, Dogri and Gojri broadcasts from Srinagar and Jammu while the Balochi broadcasts from Radio Kabul were monitored by the Quetta Unit. A small monitoring unit was established at Dhaka in 1971. In the closing days of the year 1971 when all connections between East and West Pakistan had been broken, the Monitoring Units in the two wings were the only link available. At that time there were about 45 monitors who monitored twenty stations in 15 languages and a number of Reports were brought out viz. five sets of daily Monitoring Report, Selected Items Report, Significant Items Report, Weekly Analysis, Fortnightly Report and Monthly Report.

After incorporation of Radio Pakistan, the Directorate of Monitoring was merged with CNO, reorganized and later shifted to Rawalpindi in 1975. The size of the operation of the Monitoring Unit has been reduced and there are about 24 Monitors, full- and part-time. There are only two regular Monitors against the sanctioned strength of 55. At present only one Monitoring Report and one Weekly Analysis are brought out. A special summary is also prepared for official visits of the President or Prime Minister. Since December, 1997 the Monitoring Reports are prepared on computer and are issued daily in a booklet form. The computerized archiving of these Reports is helpful in researching the old manuscripts.

Besides this Monitoring Department, CNO has always had a small News Listener Unit from the very beginning. Initially, it was also manned by persons who opted to live in Pakistan after Partition. Its purpose is to monitor English news bulletins of selected stations of the world and report immediately to the General News Room. The transcript is available within minutes of the broadcast. If there is anything special or urgent it is flashed immediately. This is a vital and quick source of news and comparison for the CNO's editorial staff. This Unit was headed by a Chief News Listener. The News Listener Unit monitored 21 daily English news bulletins of a total duration of 210 minutes. These include 14 bulletins from BBC, six of AIR and one news bulletin of Radio Dubai. The nit is manned by three Chief News Listeners and eight News Listeners.

The Monitoring Unit and the NL Unit together monitor 70 broadcasts for nearly 1800 minutes daily from 20 foreign stations in six languages (English, Urdu, Pushto, Sindhi, Arabic and Persian). The stations include: BBC, VOA, VOG, AIR, Australia, Netherlands, Switzerland, Moscow, Tehran, Beijing, Japan, Bangladesh, Tashkent, Makkah, Kuwait, Jordan, Dubai, Israel, Srinagar and Kabul.

TV Monitoring

The unit to monitor TV channels was started in May 1998 with just one TV set. Later it was strengthened and a full-fledged unit created. It had serious staffing problems in the beginning with incompetent staff. The TV Monitoring Unit is manned by retired persons who had worked in the Monitoring Unit of CNO.

At present five TV channels i.e. BBC, CNN, Zee TV, Doordarshan and Star Plus in English and Hindi are being monitored. Twenty seven (BBC 11, CNN 8, Zee 5 and DD 3) TV news bulletins with a duration of 1335 minutes are monitored daily. Besides news and commentaries, the Unit also monitors TV talks and speeches.

Computerization

To avail of the latest information technology, PBC embarked on a programme of computerization, beginning with the News Wing in 1996. In the late sixties and seventies computerizing of the inventory in the Engineering Wing had begun. Despite financial and other constraints CNO has been fully computerized. Under this system news is received by computer instead of teleprinter, it is edited on the computer and is fed direct to the studio where it is read by the newsreader from the computer screen.

Computerized English and Urdu news bulletins were on the air for the first time on 14 August 1997, the day of the Golden Jubilee of Pakistan. The national English and Urdu bulletins were transferred to Internet both in text and sound. Internet users can access this at website www.radio.gov.pk. A year later radio programmes were also linked with Internet. The changeover became effective from 12 November 1998 from 6.15 a.m. to 12 midnight. For the present, programmes of Islamabad Station along with some programmes of Rawalpindi Station, World Service and FM 101 have been linked. Pakistanis living in foreign countries can benefit from this service even without owning a transistor set. There is also no need for locating different frequencies. One can get the text as well as the voice of the newsreader through just one command.

Software is prepared for the News Wing for five types of work: news bulletins, radio monitoring reports, current affairs programmes, External Services and storage and retrieval of news records. This has been connected with the Main Server and the network consists of 50 computers. Internet and e-mail have replaced teleprinter and telex so that messages can be exchanged between cities within and outside the country very quickly and cheaply. Despite its global reach some doubt the viability and advisability of Internet for such reasons as: it is still in its infancy; a person can benefit only if he has a computer and a telephone connection to the Internet; it is not a broadcast system (i.e. one-to-many) but a telecommunication system (i.e. one-to-one); if the tariff is on the amount of data transmitted the cost goes up with the increase in the number of users.

The second phase of computerization envisages receipt of voice-reports and actualities from all over Pakistan through computers. These will be edited and broadcast in order to improve the voice quality and variety in newsreading.

Computers have also been provided to the news units at provincial headquarters of Karachi, Lahore, Peshawar and Quetta. They are linked with the Central News Organization in Islamabad through networking to ensure speedy and timely transmission of news. For example, Karachi News Unit was linked on 3 May 2000. This also provided Provincial Units access to the centralized news gathering sources at Islamabad.

INFORMATION TECHNOLOGY UNIT

PBC has decided to establish a separate information technology unit to cater to the changing technological needs of broadcasting. This unit is to play an instrumental role in the automation and computerization of radio news, programme production and editing, digitalization of its archives and audio and text transfer of PBC programmes to other broadcasters and users.

MUMTAZ HASAN BROADCASTING COMMITTEE RECOMMENDATIONS

On the subject of news and current affairs, the Mumtaz Hasan Broadcasting Committee made very far-reaching and important comments and recommendations. In its report submitted to the Government the Committee says that despite the fact that Radio Pakistan is a State-controlled organization, it has been objective in the presentation of news. That does not, however, imply that Radio Pakistan has always been as objective as it should have been. Most witnesses appearing before the Committee, particularly from the urban areas, complained that Radio Pakistan was generally hesitant in presenting facts as they are and had a tendency to play it safe. The Committee felt that truthful reporting of facts should be adhered to at all times.

Referring to a policy directive issued to Radio Pakistan the Committee said:

> We find it difficult to understand why Radio Pakistan in the presence of these unequivocal instructions to be fair, objective and factually up to date, and despite its record, has not always been able to enjoy the complete confidence of its listeners.

In the opinion of the Committee the general feeling of the listeners is that news bulletins are generally loaded with unimportant statements of Ministers, tour itineraries of VIPs and long narrations of facts and figures relating to development projects. This has resulted in a shifting of the emphasis from hard news.

The Committee considered the present timings, duration and frequency of news bulletins inadequate and unsatisfactory. The practice of having three morning bulletins in Bengali, Urdu and English one after the other, needs to be modified. It recommended that news bulletins should be spaced properly and its duration should be decreased. It said that the bulletins do not give adequate coverage to international events and insignificant local news items is sometimes repeated in more than one bulletin.

The Committee recommended that there should be a news bulletin every hour in accordance with the practice of major radio networks all over the world. The Committee in its report laid down a time schedule for hourly bulletins from 6 a.m. till midnight in English, Bengali, and Urdu. It suggested that three additional daily news bulletins in Punjabi should be introduced from Lahore and Rawalpindi to give Punjabi equal time with other regional broadcast languages.

The Committee also suggested that Radio Pakistan should have correspondents at important district towns. To ensure adequate and dependable coverage of foreign news, special correspondents should be posted abroad. The Associated Press of Pakistan and Radio Pakistan resources could be pooled and correspondents posted in major capitals of the world.

10

PLANNED DEVELOPMENT (ENGINEERING)

At the time of Independence, Pakistan Broadcasting Service inherited three small medium wave stations that could cover 6.3 per cent and 21 per cent of area and population respectively in West Pakistan and 7.8 per cent of the area and 14.8 per cent of the population in East Pakistan. On an all-Pakistan basis this meant coverage for 6.8 per cent of area and 15.7 per cent of the population. The Broadcasting Houses and Receiving Centres at Lahore and Dhaka were functioning in rented buildings. There was no national service nor any centralized newscasts. This meant that the broadcasting organization in the country had to be re-established from scratch.

Today PBC has an integrated broadcasting system consisting of high power medium wave and short wave transmitters for Home and External Broadcasts. It has FM broadcasting on a limited scale for sophisticated listeners and there are plans to expand this service in the future. This progress has been possible through the determined effort, hard work and dedication of the engineers of Radio Pakistan, who persevered in the face of great odds, a shortage of funds, frustrating red tape and many other difficulties. Also to their credit is the fact that the entire planning, development, installation, operation and maintenance of the sophisticated and complex electronic equipment was carried out by Pakistani engineers without the assistance of foreign technical experts. Radio Pakistan is perhaps the only Government department that can claim this achievement.

Pakistan's broadcasting service was developed with certain basic considerations in mind. Primary among them was that the entire country should receive medium wave service and direct broadcasts of national programmes and news from the capital station. External Services should broadcast with a strong signal from the capital station for good reception in the target areas, and technical facilities should be provided for second channel educational and religious programmes in areas required. FM broadcasting should be introduced in the country.

Priority Programme of Development—1948-55

A priority programme of technical development was launched immediately after Independence. It was divided into two phases. The first phase covered the period up to 1951 and the second phase from 1952 to 1955. The first phase had several aims: to counter Indian propaganda against Pakistan stemming from their invasion of Kashmir; to provide a service for the Federal Capital; to provide effective link between the two wings of the country; and to set up the External Services to give Pakistan a voice in the world. The second phase aimed to set up four zonal or regional centres to relay services to stations in both flanks of the country. This phase could not be achieved due to shortage of funds. The following transmitters were installed and commissioned under this programme:

Year	Station	Transmitter/Achievement
1948	Rawalpindi	500 watt SW. Later called Rawalpindi-III.
	Karachi	100 watt MW
	Karachi	10 KW MW
1949	Dhaka	7.5 KW SW
	Lahore	1 KW SW
	Karachi	50 KW SW Two Nos.*
1950	Rawalpindi	1 KW MW
1951	Karachi	Broadcasting House
	Karachi	Emergency Monitoring Centre.
1953	Rawalpindi	10 KW MW (Replaced 1 KW MW)
	Karachi	10 KW SW
1954	Karachi	10 KW SW
	Chittagong (Relay)	1 KW MW
	Rajshahi (Relay)	1 KW MW
1955	Hyderabad	1 KW MW

* Along with these short wave transmitters also provided were emergency studios, power generating plant because of lack of power connection from the local electric company, HF aerial system and residential colony for the staff of HPT Landhi.

The priority programme was successful in putting Pakistan on the map of the world broadcasting network. With the installation of four External Services on 14 August 1949, Pakistan's voice could be heard in the Middle East, Iran, Turkey, Afghanistan, South East Asia, Far East, England and parts of Europe. National service from Karachi could be heard all over the country on short waves. Regional medium wave services could be broadcast from major towns in West and East Pakistan. A zonal short wave service was started from Lahore. Total radiated power increased from 20 KW to 172.5 KW. Offices of HPT Landhi, Directorate of News and Radio Stations at Karachi and Rawalpindi were added to the set-up. A Central Stores Depot was constructed at Landhi for proper storage of equipment.

First Five Year Plan—1955-60

The First Five Year Plan of technical development was formulated to increase coverage in West and East Pakistan on medium wave and to supplement it with zonal services on short wave. During the Plan period a 1 KW MW transmitter was set up at Quetta, a 10 KW MW transmitter at Hyderabad, a 10 KW SW transmitter each at Dhaka and Peshawar, an international Monitoring Centre at Karachi and a Broadcasting House at Dhaka. The overall position of broadcast transmitters at the end of the First Five Year Plan was as under:

Station	Short Wave Transmitter	Medium Wave Transmitter	Studios
Karachi	2-50 KW		
	2-10 KW	1-10 KW	Broadcasting House
Lahore	1-1 KW	1-5 KW	Studios in hired building.
Hyderabad	—	1-10 KW	—do—
Quetta	—	1-1 KW	—do—
Rawalpindi	1-10 KW	1-10 KW	Limited production facilities.
Dhaka	1-7.5 KW		
	1-10 KW	1-5 KW	Broadcasting House
Peshawar	1-10 KW	1-10 KW	Broadcasting House with few studios.
Chittagong	—	1-1 KW	Relaying Station.
Rajshahi	—	1-1 KW	—do—
Total	9-158.5 KW	9-53 KW	

SECOND FIVE YEAR PLAN—1960-65

The following technical facilities were added during the Second Five Year Plan:

Year	Station	Facility/Transmitter
1960	Rawalpindi	Additional Studios
	Karachi	Receiving Centre
		Peshawar Receiving Centre
1961	Quetta	10 KW MW Transmitter
	Sylhet	2 KW MW Transmitter
	Quetta	Receiving Centre
1962		Chittagong 10 KW MW Transmitter
	Lahore	Receiving Centre
	Rajshahi	10 KW MW Transmitter
	Quetta	10 KW SW Transmitter
	Hyderabad.	Broadcasting House
1963	Dhaka	Receiving Centre
	Dhaka.	100 KW MW Transmitter
	Dhaka	Extension to Broadcasting House
1964	Islamabad	Receiving Centre
	Chittagong	Broadcasting House
	Rajshahi	Broadcasting House
	Quetta	Broadcating House
1965	Lahore	100 KW MW Transmitter
	Lahore	Broadcasting House

The position of broadcast transmitters at the end of the Second Five Year Plan was as under:

Station	Short Wave Transmitter	Medium Wave Transmitter	Studios
Karachi	2-50 KW		
	2-10 KW	1-10 KW	Broadcasting House
Hyderabad	———	1-10 KW	Broadcasting House
Lahore	———	1-100 KW	
		1-5 KW	Broadcasting House
Rawalpindi	1-10 KW	1-10 KW	Limited studios.
Peshawar	1-10 KW	1-10 KW	Broadcasting House
Quetta	1-10 KW	1-10 KW	Broadcasting House
Dhaka	1-10 KW		

	1-7.5 KW	1-100 KW	
		1-5 KW	Broadcasting House
Chittagong	————	1-10 KW	Broadcasting House
Rajshahi	————	1-10 KW	Broadcasting House
Sylhet	————	1-2 KW	Limited production facilities.
Total	9-167.5 KW	12-282 KW	

Achievements at the end of the Second Five Year Plan were:

1. The number of transmitters increased from 3 to 21.
2. The radiated power increased from 20 KW to 449.5 KW.
3. The number of Broadcasting Houses increased from 3 to 10.
4. The number of Receiving Centres increased from 2 to 5.
5. Area and population coverage on medium wave in East Pakistan increased from 8 per cent and 15 per cent to 95 per cent and 96 per cent respectively.
6. Area and population coverage on medium wave in West Pakistan increased from 6 per cent and 21 per cent to 25 per cent and 62 per cent respectively.

THIRD FIVE YEAR PLAN—1965-70

The broad objectives of the Third Five Year Plan in the field of broadcasting were:

1. To cover both Wings of the country with reliable and effective medium wave services.
2. To establish strong inter-wing links on short waves.
3. To extend the external services both in duration and in target areas by increasing the number and power of short wave transmitters.
4. To replace old and worn out facilities with more modern and sophisticated equipment.

The following technical facilities were added during the Third Five Year Plan:

Year	Station	Transmitter/Facility
1967	Rangpur	1-10 KW MW Transmitters
1968	Islamabad	2-100 KW SW Transmitters
	Islamabad	1-10 KW SW Transmitter
	Rawalpindi	1-100 KW SW Transmitter
	Dhaka	1-100 KW SW Transmitter
1969	Karachi	1-1 KW MW Transmitter
1970	Islamabad	Short wave aerial system

The medium wave coverage in West Pakistan did not increase to any sizeable extent during this period. However, a second channel was added on medium wave at Karachi. Short wave power in West Pakistan increased by 310 KW. Inter-wing links were established on 100 KW SW transmitters from Islamabad and Dhaka. The total capital investment up to the end of the Third Five Year Plan was Rs. 135.5 million.

Fourth Five Year Plan—1970-75

The objectives of the Fourth Five Year Plan in the field of broadcasting development were:

1. To extend and strengthen the medium wave coverage in the country.
2. To provide second broadcast channels at all major stations in the country.
3. To further strengthen and extend the external coverage.
4. To further strengthen the inter-wing broadcast link.
5. To strengthen and provide more studio facilities.
6. To provide short wave transmitters at Dhaka for External Services.

The Fourth Five Year Plan included the following major projects:

West Pakistan

On-going Projects
1. 2-250 KW SW Transmitters, Islamabad.
2. Broadcasting House, Islamabad, Peshawar and Multan.
3. 1000 KW MW Transmitter, Islamabad.

PLANNED DEVELOPMENT (ENGINEERING)

4. 2-100 KW SW Transmitters, Islamabad.
5. 150 KW MW Transmitter, Quetta.
6. 120 KW MW Transmitter each at Hyderabad and Multan and Receiving Centre, Multan.
7. 2 FM and 2 MW Transmitters, Karachi and Lahore.
8. 10 KW MW Transmitter, Khairpur.
9. Modernization.

New Projects
1. Residential Colony, Islamabad.
2. 100 KW MW Transmitter, Peshawar.
3. Residential accommodation at Stations for some essential staff.
4. Mobile Transmitters.
5. Broadcasting House, Karachi.
6. Extension Broadcasting House, Hyderabad and Quetta.

EAST PAKISTAN

On-going Projects
1. 1000 KW MW Transmitter, Dhaka.
2. 1-100 KW SW Transmitter and aerials, Dhaka.
3. 1-10 KW MW Transmitter, Khulna.
4. 1-FM and 1-10 KW MW Transmitter, Dhaka.
5. 100 KW MW Transmitter, Khulna.
6. Broadcasting House, Khulna.
7. Extension Broadcasting House, Dhaka.
8. Modernization.

New Projects
1. Broadcasting House at 2nd Capital, Dhaka.
2. 2-100 KW SW Transmitters, Dhaka.
3. Technical Services.
4. 100 KW MW Transmitter each at Chittagong and Rajshahi.
5. 10 KW MW Transmitter, Sylhet.
6. Broadcasting House, Rangpur and Sylhet.
7. Residential colony at stations and at 100 KW Transmitter, Dhaka.
8. Mobile Transmitters.
9. Recording Studio at Sylhet, Rangpur and Khulna.
10. Extension Broadcasting House, Chittagong and Rajshahi.
11. 100 KW MW Mast Installation and Civil works at Dhaka.

The following facilities were added by December 1971:

Year	Station	Transmitter/Facility
1970	Lahore	Transcription Service Studios
	Islamabad	Technical Training School and Staff Training School
	Multan	120 KW MW Transmitter
	Khulna	10 KW MW Transmitter
1971	Hyderabad	120 KW MW Transmitter

Due to civil war and Indian aggression in East Pakistan in December 1971 it was not possible for Radio Pakistan to implement the expansion programme there. The position in March 1972 in West Pakistan (now Pakistan) was as given below:

Station	Short Wave Transmitter	Medium Wave Transmitter	Studios
Karachi	2-50 KW	1-10 KW	Broadcasting House
	2-10 KW	1-1 KW	
Hyderabad	——	1-120 KW	Broadcasting House
		1-10 KW	
Quetta	1-10 KW	1-10 KW	Broadcasting House
Multan	——	1-120 KW	Limited Studio Facilities
Lahore	——	1-100 KW	Broadcasting House
		1-5 KW	
Rawalpindi	1-100 KW	1-10 KW	Broadcasting House
	1-10 KW		
Islamabad	2-100 KW	——	——
	1-10 KW		
Peshawar	1-10 KW	1-10 KW	Old Broadcasting House
Total	11-460 KW	10-396 KW	

Following is the comparative operational position as at the time of Partition, at the end of the Second and Third Five Year Plans, in December 1971 and in March 1972:

	1947	End of 2nd Plan 1965	End of 3rd Plan 1970	December 1971	West Pakistan March 1972
No. of Transmitters	3	22	28	31	21
MW	3	12	14	17	10
SW	-	10	14	14	11
Power of Transmitters	20	449.5	870.5	1120.5	856
KW MW	20	282.0	293.0	543.0	396
KW SW	-	167.5	577.5	577.5	460
No. of Broadcasting Houses	3	12	13	15	9
No. of Receiving Centres	2	6	6	6	5
Coverage in EP					
Area (%)	8	95	98	100	—
Population (%)	15	96	99	100	—
Coverage in WP					
Area (%)	6	25	25	33	33
Population (%)	21	62	62	78	78

CRASH PROGRAMME—1972-75

The expansion programme suffered due to the political situation in the country during the period 1970-72. After the dismemberment of the country the development plan for the remaining period of the Fourth Five Year Plan had to be reviewed and priorities reset. A priority or crash programme was framed which included almost all the on-going projects to fill in the gaps in the medium wave coverage. The aims and objects of the crash programme were:

i) To extend and strengthen the medium wave coverage in West Pakistan.
ii) To transmit information about the reforms, policies and activities of the Government to every corner of the country.
iii) To provide second broadcast channels at all the major stations.
iv) To radiate external services with a strong signal from Islamabad to create good reception conditions in the target areas.
v) To introduce more sophisticated third channel broadcasts on Frequency Modulated Transmitters from all major stations.

The following major projects were included in the Crash Programme:

1) 2-250 KW SW Transmitters, Islamabad.
2) 2-100 KW SW Transmitters, Islamabad.
3) Broadcasting Houses, Islamabad, Peshawar and Multan.
4) 150 KW MW Transmitter, Quetta.
5) 1000 KW MW Transmitter, Islamabad.
6) Extension of Broadcasting House and Transcription Service, Karachi, Central Store Building, Landhi.
7) Receiving Centre, Hyderabad.

The following projects were completed during the period from 1972 to 1975:

1972	2-250 KW	SW Transmitters, Islamabad.
1973	1-150 KW	MW Transmitter, Quetta.
1974	2-100 KW	SW Transmitters, Islamabad.
1975	1-10 KW	MW Transmitter, Bahawalpur.
	1-50 KW	MW Transmitter, Lahore.

TWENTY YEAR DEVELOPMENT PLAN—1975-95

A development plan of broadcasting in Pakistan for the period 1975-95 was prepared with the following objectives:

1) The entire country gets good medium wave coverage.
2) Programmes of national importance and news originating from Islamabad reach every nook and corner of the country.
3) Programme origination facilities are increased at the existing Broadcasting Houses.
4) Auditoriums are added to all the main Broadcasting Stations to provide facilities for audience-participated programmes.
5) Technical facilities are provided for second channel programmes from all the main stations.
6) External Services are radiated on a stronger signal and to a larger number of targets both on medium wave and short wave transmitters.
7) Added facilities are provided in the fields of training, monitoring and research.

8) Old and worn out equipment is replaced at old transmitting stations, broadcasting houses and receiving centres with new and modern equipment.
9) More sophisticated frequency modulation technique is introduced to provide quality reception as a third channel from all the main broadcasting stations.

FIFTH FIVE YEAR PLAN—1975-80 (LATER 1978-83)

Keeping in view the broad guidelines of the long-term development plan, the Fifth Five Year Plan was formulated with the following objectives:

1. To use radio as a means of socio-economic uplift of masses by strengthening the development support communication and enhancing appeal of other broadcasts.
2. To create physical facilities for the extension of medium wave radio coverage to 72 per cent of area and 96 per cent of population by the end of the Plan period.
3. To improve the financial viability of the Corporation through
    a) a better system of collection of licence fees, and
    b) increased revenues through commercial advertisements, etc.
4. To provide
    a) second channel at all stations to relieve congestion in the main network; and
    b) to devote more time to regional and special audience programmes.
5. To enhance the educational and instructional role of radio in cooperation with the Open University.
6. To improve the existing facilities through modernization of equipment.
7. To undertake programmes aimed at local assembly of transmitting equipment, including FM Transmitters, up to 100 KW Transmitters, master switching units and sound equipment to save on foreign exchange cost of imports.
8. To explore the possibility of manufacture of certain components of receiving sets in the country.
9. To devise suitable methods and machinery for the evaluation of programmes.

The Fifth Plan was initially for the period 1975-80 but was later extended to the period 1978-83. The following technical facilities were added during the three year period from July 1975 to June 1978:

1976	10 KW	MW Transmitter, Rawalpindi-II.
1977	100 KW	MW Transmitter, Karachi.
	300 KW	MW Transmitter, Peshawar.
	Inauguration of Islamabad Station.	
	1000 KW	MW Transmitter, Islamabad.

The total radiated power by the end of June 1978 increased to 2011 kilowatts on medium wave and 1161 kilowatts on short wave through 32 transmitters. The area coverage on medium wave increased to 59 per cent and population coverage rose to 88 per cent. The following facilities and installations were completed up to the end of the Fifth Five Year Plan period, i.e. June 1983:

1979	0.25 KW	MW Transmitter, Gilgit.
	0.25 KW	MW Transmitter, Skardu.
1981	0.25 KW	MW Transmitter, Turbat.
	10 KW	MW Transmitter, D.I.Khan
	100 KW	MW Transmitter, Khairpur.
	0.25 KW	MW Transmitter, Khuzdar.
1982	10 KW	MW Transmitter, Skardu.
1983	100 KW	MW Transmitter, Rawalpindi-II.

SIXTH FIVE YEAR PLAN—1983-88

At the end of the Fifth Five Year Plan on 30 June 1983 Pakistan Broadcasting Corporation was providing medium wave coverage to 62 per cent of the area and 91 per cent of the population in the country. There were some wide area gaps that were not covered by PBC signal. These gaps were mostly in Balochistan, Northern Areas and some pockets in other Provinces. The Sixth Five Plan when initially drafted, included a large number of schemes which were supposed to fill these wide gaps. These included a 300 kilowatt medium wave transmitter at Khuzdar and a large number of 10 kilowatt stations. Some short wave transmitters were included to provide coverage within the country and also for the use of External Services.

PBC had also planned a network of FM service for implementation in two phases. The first phase was included in the Sixth Plan comprising seven transmitters located at important places in the country which would have provided sufficiently wide coverage of quality broadcasts on FM. Due to financial constraints a sum of Rs.370 million was allocated to PBC in the Sixth Plan period as against the demand of Rs.1132 million. According to the revised plan it was to cover 89 per cent of the area and 96 per cent of the population by the end of June 1988. The actual allocation was almost half of this amount. The progress was, therefore, not as planned and a number of schemes were carried over to the Seventh Plan.

Twenty Year Perspective Plan—1985-2005

In the meantime a twenty year perspective plan was prepared for the development of broadcasting in the country, with the objectives as outlined below:

1. Replacing the old equipment and providing modern equipment at various units.
2. Providing a second channel at the provincial capitals and other main cities of the country for alternate programmes for the purpose of education, sports, agricultural support and entertainment.
3. Providing external services on short wave for America, Canada and other major targets in the world.
4. Replacing old short wave transmitters with new equipment.
5. Transfer of sound archives on to compact discs.
6. Providing sound broadcasting on FM (mono and stereo) for sophisticated listeners in the country.
7. Replacing old aerial systems in poor condition.
8. Providing a housing colony for employees at big cities.
9. Providing first grade signal of sound broadcasting on medium wave to 100 per cent of the population and area in the country.

This perspective programme of broadcasting development is to be phased according to the budget allocations given by the Government during various Plan periods.

Seventh Five Year Plan—1988-93

The Seventh Five Year Plan for sound broadcasting was prepared keeping the following objectives in view:

1. To extend interference free medium wave coverage to 98 per cent of the population and 92 per cent of area in the country.
2. To replace the old 10 kilowatt transmitters with 100 KW MW transmitters at Rawalpindi and Quetta for second channel operation.
3. (a) To improve the signal of external services by providing high power short wave transmitters.
   (b) To introduce new external services for SAARC countries and other countries which are presently not fed by any external service.
   (c) To strengthen the World Service for the benefit of Pakistanis living abroad.
4. To replace and modify old studios and transmitting equipment at the existing units which have outlived their useful life.
5. To improve studio facilities at some of the existing broadcasting houses having inadequate facilities.
6. To introduce a high quality sound broadcasting service on FM to cater to the need of quality conscious listeners.

In order to achieve the above-mentioned objectives the following schemes were planned for the Seventh Five Year Plan:

On-going Schemes
1. Broadcasting Houses, Islamabad and Karachi.
2. 300 KW MW Transmitter and BH Khuzdar.
3. 100 KW MW Transmitters, Peshawar and Karachi.
4. 10 KW MW Transmitters and BHs, Faisalabad, Gilgit, Skardu, Loralai and Zhob.
5. Village Broadcasters, Sibi, Abbottabad, Rahim Yar Khan, Mirpurkhas, Chitral and D.G. Khan.
6. Balancing and modernization of equipment, phase-III,
7. 2-250 KW SW Transmitters, Islamabad.
8. Additional security works.
9. Minor Projects.
10. Modification of Transmitters.

PLANNED DEVELOPMENT (ENGINEERING)

11. Replacement of damaged aerial system at Rewat.
12. Housing Scheme (Phase-I) for PBC employees at Sector G-8/4, Islamabad.

New Schemes
1. Balancing and modernization of equipment, phase-IV.
2. 100 KW MW Transmitters, Quetta and Rawalpindi.
3. 2-250 KW SW Transmitters for External Services.
4. Transfer of sound archives on to compact discs.
5. Extension Broadcating House, Hyderabad.

Against a demand of Rs.1903 million for the implementation of the above-mentioned on-going and new schemes in the Seventh Plan an amount of Rs.221 million was finally allocated. Consequently many on-going schemes could not be completed during Plan period and were carried over to the Eighth Five Year Plan.

EIGHTH FIVE YEAR PLAN: 1993-98

The Eighth Five Year Plan for radio had the following objectives in view:

1. To make the radio signal on medium wave band cover 100 per cent of the population and 90 per cent of the area of the country.
2. To strengthen the coverage of short wave signal with the help of powerful short wave transmitters and to improve the reception of the World and External Services of PBC.
3. To improve the reception and coverage area by replacement of old and obsolete 10 KW MW transmitters at Peshawar, Rawalpindi, Karachi and Quetta by 100 KW MW transmitters.
4. To improve the signal of PBC on short wave band by replacing old and obsolete short wave transmitters with new powerful transmitters.
5. To start additional external services for countries which are not covered presently.
6. To improve the broadcast quality in accordance with listeners' expectations by replacing the old studios/transmitting equipment.

7. To modernize the Research and Development Section of Equipment Production Unit and to digitalize the broadcast equipment.
8. To help solve the residential problems of low paid employees with the construction of residential colonies at new projects.
9. To create employment opportunities through the establishment of new stations and HPTs.

To fulfil these objectives the following schemes were drawn up for the Eighth Five Year Plan:

On-going Schemes
1. Broadcasting House, Karachi.
2. 100 KW MW Transmitter each at Karachi, Peshawar, and Rawalpindi.
3. 10 KW MW Transmitter and BH, Loralai and Zhob.
4. 100 KW MW and 100 KW SW Transmitters and BH Mirpur.
5. Balancing and modernization of equipment, phase-IV.
6. Additional security works.

New Schemes
1. 2-250 KW SW Transmitters for External Services at Rewat, Islamabad.
2. Transfer of sound archives on to compact discs.
3. Replacement of radio equipment and BH, Islamabad, Lahore and Multan.
4. Replacement of 2-50 KW SW Transmitters and aerial system with 2-100 KW SW Transmitters, Karachi.
5. Replacement of 2-100 KW SW ampliphase Transmitters, Rewat, Islamabad.
6. 4-100 KW SW Transmitters and aerial system, Islamabad.
7. 1000 KW MW Transmitter for Middle East and Gulf at Gwadar.
8. Minor Projects.

The Eighth Five Year Plan was approved with the allocation of Rs.296.3 million for on-going projects and Rs.1087.0 million for new schemes, i.e. a total of Rs.1383.3 million. Among the schemes included in the Eighth Five Year Plan, only three were completed namely 10 KW MW Transmitter and BH each at Loralai and Zhob and 100 KW

MW Transmitter, Peshawar. Other schemes could not be completed due to shortage of funds.

NINTH FIVE YEAR PLAN

The Ninth Five Year Plan is under process. It includes a number of important schemes besides the on-going projects carried over from the last Plan. On the completion of the schemes included in the Ninth Plan a marked improvement is expected in the signal strength and quality of programmes. The establishment of new radio stations on medium wave will cover those far-flung areas of the country that are not presently covered, such as Gwadar. Some SAARC countries and states of Central Asia and North Africa presently not covered by short wave signal, will be reached by the installation of new short wave transmitters. The duration of existing World/External Services will be enhanced for the benefit particularly of expatriate Pakistanis living in the target areas. No short wave transmitter has been added since 1974.

FUTURE PLAN

Pakistan Broadcasting Corporation has drawn up a plan for the development of sound broadcasting in the future. Schemes included therein will form part of the coming Five Year Plans. Some of the projects included in the future plan are given below:

1. 2-250 KW SW Transmitters for External Services at Islamabad.
2. 2-250 KW SW Transmitters, Karachi.
3. 2-100 KW SW Transmitters, Landhi, Karachi (Replacement of 2-50 KW SW Transmitters).
4. 4-100 KW SW Transmitters, Islamabad (Replacement of existing 4-100 KW SW Transmitters).
5. 1000 KW MW Transmitter and Broadcasting House at Gwadar.
6. 100 KW MW Transmitter and BH each at Rahim Yar Khan and Panjgur.
7. 100 KW MW Transmitter at Quetta (Replacement of 10 KW MW Transmitter).
8. Upgrading of Village Broadcaster to 10 KW MW Station at Turbat, Sibi, Abbottabad and Chitral.

9. Replacement of old and obsolete radio equipment at different stations/units.
10. 500 KW MW Transmitter each at Hyderabad and Lahore.
11. Establishment of FM network in the country.
12. Coverage of coastal highway by FM Service.
13. 100 KW SW Transmitter and aerial system at Skardu.
14. Introduction of Digital Audio Broadcast from Islamabad, Karachi, Lahore, Peshawar and Quetta.
15. Modernization of Research and Development Section of EPU for digitalization of radio equipment.
16. Minor Projects.

FM BROADCASTING

FM transmissions are carried out on the VHF band. These waves travel along the line of sight similar to television. The transmission can be heard clearly within a radius of fifty to sixty-five kilometres. Repeater transmitters are needed after every hundred kilometres or so to provide national coverage. The advantages of FM transmission are, it is economical in operation as the power of transmitter is very small, it always provides good quality programmes, is used as stereo transmission, and is free from jamming.

In Pakistan, major regional stations like Peshawar, Islamabad, Lahore, Multan, Hyderabad, Karachi and Quetta have FM transmitters. These carry the same programmes as the local medium wave transmitter or they serve as a stand-by studio-transmitter link. This is not the ideal use of FM transmitters.

The spectrum of medium wave transmission is becoming more and more congested with the increase in high-power transmitters. This congestion in the spectrum has resulted in severe interference. To give listeners what they want, particularly in the field of music, as well as sports, city service, news and current affairs the superior sound quality of FM service is necessary. The system is already used by most of the advanced broadcasting organizations in the world.

FM or VHF links are required to supplement the non-exchange telephone lines between Receiving Centres, Broadcasting Houses and Transmitters. The necessity of these STLs has arisen because of the unsatisfactory performance of telephone lines and frequent, often long, interruptions in the programmes due to faults on the non-exchange telephone lines. These links improve the quality of broadcasts and ensure uninterrupted service.

Satellite Communication

Since the establishment of Radio Pakistan the news, national hook-up and some other programmes have been rebroadcast or relayed by the stations after receiving them on professional communication receivers. These relayed programmes contained propagational noise, hum, fading, and sometimes cross talk of a nearby powerful station due to congestion in the short wave band. This marred the quality of relayed programmes. In order to overcome this problem these programmes were fed to different stations of PBC through non-exchange telephone lines. The quality of programmes improved to some extent but was still not clear due to the performance of long distance telephone lines. Some PBC stations situated in far-flung areas had to depend on receivers to relay the news and national hook-up programmes.

PBC, however, continued its efforts to improve the quality of relayed programmes. In July 1994 it took a big step towards progress by venturing into satellite communication with the cooperation of Pakistan Television Corporation. PBC acquired a slit of 200 kHz in Transponder hired by PTV in Asiasat and established the system to broadcast programmes of Islamabad Station, national and regional news and hook-up programmes on this link. All the PBC Stations were provided with necessary equipment for the reception and rebroadcast of these programmes.

Due to this new system the quality of relayed programmes became clear. PBC thus successfully overcame the long outstanding problem of quality of relayed programmes. The programmes of Islamabad Station can now be heard throughout Pakistan and in 36 other countries of the world through a dish antenna for about 18 hours from early morning to midnight i.e. from 0545 hrs to 2400 hrs PST (0045 hrs to 1900 hrs UTC).

Mobile Recording Van

It is difficult for a broadcasting organization to function efficiently unless facilities for outside broadcasts are easily available. The need for such facilities was growing day by day. Telephone lines, in certain cases, were not available at all with the result that excellent programmes had to be sacrificed for want of a mobile unit that could be transported to the spot for actual recording. Mobile recording vans were the answer to this problem. These vans could only be effective if they were self-contained.

Radio Pakistan engineers designed a Mobile Recording Van which had two tape recorders for continuous recording and these were operated by a converter run on accumulators. A petrol generating plant was also mounted within the van that could be used for charging the accumulators where power supply was not available, for example, as in villages remotely located from the highway. Two bunks were also provided for the staff who were expected to spend a few days away from their Headquarters. Some vans were imported. With the availability of portable recording gear these vans gradually went out of use.

UNESCO Experts

Four UNESCO experts visited Radio Pakistan under Technical Assistance Programmes. One expert on school broadcasts remained in Pakistan for about two and a half years. On his recommendations a School Broadcast Unit was set up at Karachi Station in 1952 and School Broadcast started in early 1953.

The remaining three experts included two Field Work Engineers for carrying out 'field measurement' for the purpose of determining conductivity of soil, noise level, field intensity and collection of other essential data, and one Studio Expert. The Studio Expert was assigned the job of designing the proposed Broadcasting House at Dhaka. He returned home before the plans were ready. The Field Work engineers stayed on for over a year.

Frequency Management

PBC has in all 50 medium wave frequencies registered with the International Telecommunication Union (ITU), Geneva. Out of these registered frequencies the following are in use from various stations with various powers:

PLANNED DEVELOPMENT (ENGINEERING)

Station	Frequency (kHz)	Power (KW)
Peshawar-I	540	300
Khuzdar	567	300
Islamabad	585	1000
Lahore-I	630	100
Karachi-II	639	10 (being replaced with 100 KW)
Peshawar-II	729	100
Quetta-I	756	150
Rawalpindi-II	792	150
Karachi-I	828	100
Quetta-II	855	10
Khairpur	927	100
Hyderabad-I	1008	120
Multan	1035	120
Lahore-II	1080	50
Hyderabad-II	1098	10
Rawalpindi	1152	10 (being replaced with 100 KW)
Loralai	1251	10
Bahawalpur	1341	10
Dera Ismail Khan	1404	10
Zhob	1449	10
Faisalabad.	1476	10
Gilgit	1512	10
Skardu	1557	10
Chitral	1584	0.25
Sibi	1584	0.25
Turbat	1584	0.25
Abbottabad	1602	0.25
	27 Transmitters	2701 KW

For short wave services (External Services and World Service) PBC has 210 frequencies registered with ITU, Geneva. Out of these frequencies 62 to 65 are usually used in the schedules for the Home, World and External Services from its transmitters as under:

HPT/Station	No. of Short Wave Transmitters	Power (KW)
1. Islamabad Complex	1 x 10	10
	5 x 100	500
	2 x 250	500
2. Landhi, Karachi.	2 x 50	100
3. Quetta.	1 x 10	10
4. Peshawar	1 x 10	10
5. Rawalpindi I & II	1 x 10	10
	1 x 1	1
Total	**14**	**1141**
Total MW & SW	41	3842
FM Transmitters	5	12
Total MW, SW, FM	46	3854

Equipment Production Unit

To overcome its perennial lack of funding problem, PBC created a new unit within its network namely the Equipment Production Unit in 1975. The aim was to save on the import of broadcasting equipment through foreign companies that were supplying transmitters but not transferring the technology, and replace worn out equipment with locally available materials.

It was expected that the Unit would supply new equipment and upgrade existing equipment with minimal foreign exchange input and great economy in overall cost. Existing worn out equipment at stations and units would be replaced with locally produced new equipment, resulting in greater uniformity in the requirement of spare parts. The standardization would allow for the production of spare parts locally as well. Fabrication of equipment in EPU was not only cost effective but also helped to achieve self-reliance in the field. Expertise in state of the art equipment and technology was an added benefit.

EPU has been fabricating transmitters of different types and wattage and scores of ancillary equipment like various types of amplifiers, consolettes, turn tables, antennae, FM links etc. The transmitters and other items produced at the Unit have been installed at various Stations and Units and are functioning satisfactorily.

PLANNED DEVELOPMENT (ENGINEERING) 151

Equipment fabricated at EPU	Quantity
1. MF Transmitters:	
i) 10 KW	3
ii) 100 KW	3
iii) 150 KW	1
iv) 300 KW	1
2. FM Transmitters:	
10/50 Watt	32
3. RF Feeder Lines for	
i) 60 Ohm for MF 500 KW Transmitter	1
ii) 150 Ohm for MF 100 KW Transmitter	3
iii) 60 Ohm for MF 300 KW Transmitter	1
iv) 300 Ohm for 250 KW SW Transmitter	2
4. T-Type Matching Networks for	
i) 150 KW MF Transmitter	1
ii) 300 KW MF Transmitter	1
iii) 500 KW MF Transmitter	1
iv) 100 KW MF Transmitter	3
5. Audio Equipment:	
i) Acoustic treatment and supply of public address system for community centres	2
ii) Audio amplifiers with 10 W/50 W	88
iii) Multi-channel audio mixing Consolettes	67
iv) Transcription turn tables along with equalizing amplifier etc.	68
6. Master Switching Matrix:	
i) With 10 inputs and 10 outputs	1
ii) With 10 inputs and 6 outputs	1
iii) With 10 inputs and 4 outputs	7
iv) With 10 inputs and 2 outputs	1
7. Standard Audio Racks	23
8. Monitor Amplifiers with wooden cabinets	40
9. FM Transmitting and Receiving Antennae	24
10. Microphone Stands boom type and table type	37
11. Conference Room Microphone System One Set of 30 Pieces	
12. Equipment under fabrication:	
i) Audio Consolettes	12
ii) Switcher 10 x 4	3
iii) FM antennae	2
iv) 10 KW MW Transmitters	2

11

RADIO AND THE NATION IN WAR AND PEACE

In the 54 years since it came into being, Radio Pakistan has faced many challenges and critical situations. Despite persistent limitations in men, material and money it has always persevered and endeavoured to provide a creditable broadcasting service for the country. Radio Pakistan's service to the nation at the time of Partition, during the Indian attack on Kashmir, and other occasions has been covered in Chapter 3. Some events of national importance covered by it since then are briefly described in the following pages.

LIAQUAT ALI KHAN'S ASSASSINATION

The news about the assassination of Liaquat Ali Khan, the first Prime Minister of Pakistan, was announced in the main evening bulletin on 16 October 1951. Regular programming ceased immediately and instead *Quran Khwani, Hamd* and *Naat*, recitation from Iqbal and *Shahnama* and other special programmes were broadcast in the spirit of mourning. The nation was devastated by his death.

Running commentary on the funeral procession and burial was broadcast on 17 October 1951 as well as on the '*soyem*' ceremony held on 19 October. Proceedings of a condolence meeting on his death were relayed from Paltan Maidan, Dhaka, including speeches of the Governor, Chief Minister, Leader of Opposition in the Provincial Legislature and other political and religious leaders. The fortnightly programme journal '*Ahang*' brought out a supplement on the occasion. The '*chehlum*' ceremony was relayed from the mausoleum of Mr Liaquat Ali Khan on 22 November 1951.

Rann of Kutch

In the national emergency that was declared during the Indo-Pakistan Rann of Kutch dispute, the complexion of Radio programmes underwent a marked change. Special informative and morale-building programmes were introduced to explain Pakistan's stand in the dispute and to effectively counter unfounded Indian propaganda. The special representative of Radio Pakistan along with foreign correspondents flew into Rann of Kutch to get on-the-spot news.

1965 War

The year 1965 was one of the most significant and memorable years in the annals of broadcasting in Pakistan. In September of that year India's aggression against Pakistan was in flagrant violation of the UN Charter. Radio Pakistan was faced with a grave challenge and responsibility of keeping the people informed. There was a spontaneous change in the objectives and content of programmes.

During the 17-day war, Radio Pakistan joined in the nation's struggle against aggression and it became known as the 'people's medium' and the 'second line of defence'. Pakistanis came to rely on Radio Pakistan for credible news reports of the war situation, rather than AIR and even BBC.

Like all other national institutions, Radio Pakistan joined the battle to defend the freedom and honour of the country. The executives, programme makers, engineers and newsmen made dedicated efforts to give accurate reports on the war. Poets, writers, scholars, composers, and artists offered their services for special war-time programming. Practically all the stations functioned round the clock. Despite blackouts and the ominous drone of enemy aircraft in the sky, they kept Radio Pakistan on the air. Radio personnel went to the battle-front and gave minute to minute accounts of the war.

Through martial music and special programmes broadcast round the clock, Radio Pakistan attempted to boost the morale of the people. Of this special programming, about 500 music and spoken-word items have been preserved in the 'War Archives' at Central Productions. The programmes produced during that war were so inspiring that the aggressor alleged that those programmes were proof that Pakistan had planned and initiated the war. Special attention was paid to the production and presentation of patriotic songs. The stations tried to evolve a new pattern of music full of vigour and rallying support for the troops. The martial and patriotic songs produced at that time are still in demand today.

Radio Pakistan's role in boosting the public morale and exposing the enemy's aggression to the world was recognized by the people, the Armed Forces and those at the helm of government. It projected Pakistan's case to the world effectively and forcefully in its home and external broadcasts. The imagination, stamina and enthusiasm of programme staff, the newsmen, the artists and engineers of Radio Pakistan galvanized the nation to a sense of patriotism, civic responsibility and awareness of the need to work hard for the socio-economic progress of the country.

Radio Pakistan began countering misleading Indian propaganda. A psychological warfare cell was set up at the start of the war in Karachi to monitor enemy broadcasts and guide the Stations and Units of Radio Pakistan in countering false reports. Due to the war situation, transmission hours and allied activities were extended. All broadcasting services and transmitters functioned uninterrupted and satisfactorily during the emergency. Despite a tremendous pressure on its resources, Radio Pakistan afforded full voice-cast facilities to war correspondents from all over the world.

An emergency news desk was set up at Rawalpindi as the bulk of war news originated from that city which was the seat of General Headquarters. The bulletins were centralized in Karachi. The news desk helped maintain a 24-hour telecommunication link between the Emergency Desk in Rawalpindi and the Central News Organization in Karachi and in close liaison with the Ministry of Information and Broadcasting, Central Information Board and GHQ. This continual link was used to guide the preparation of sixty-eight newscasts in different languages directed at listeners at home and abroad and to convey decisions on policy matters. It also facilitated a round the clock check on Indian claims and maintained a regular flow of directives for countering these claims.

Rawalpindi and Karachi Centres were also strengthened by a second teleprinter link and the duration of their operation was extended from six in the morning till midnight. The Karachi-Dhaka link was also extended from 6.00 a.m. to midnight to maintain an uninterrupted flow of news and other broadcast material to Karachi from the eastern wing and to keep the regional stations in that wing fully informed on policy matters.

There was round the clock monitoring of All India Radio's news broadcasts and the main news bulletins of BBC, VOA, Radio Moscow, Radio Beijing (then Peking), Radio Australia, Jakarta and Cairo. All references to the Indo-Pakistan war and matters relating to Pakistan in the foreign broadcasts were transmitted to the Emergency Desk in Rawalpindi for the information of the Government and other official agencies engaged in public relations.

The External Services worked hard to keep the outside world informed about the course of events. The impact of these broadcasts was to garner support and sympathy from thousands of overseas liteners for Pakistan. Letters to the External Services poured in from as far as West Africa to the Islands of Indonesia. After the war Rawalpindi Station broadcast the interviews of about 1100 Indian prisoners of war.

1971 War

Radio Pakistan's role in the 1971 war between India and Pakistan was very different from the one it played in the war of September 1965. The Stations' information sources were not always reliable and in some instances provided inaccurate details. As a result Radio Pakistan's credibility suffered considerably.

After the traumatic separation of East Pakistan, the entire nation sunk into depression. Radio Pakistan altered its programme content to emphasize nation-building and renewal of the nation's faith in its destiny. Special programmes included motivational songs, talks and features highlighting the country's potential in various fields. Major themes included the promotion of Islamic values and faith in the ideology of Pakistan, the issue of prisoners of war and displaced persons, the new Constitution and reforms introduced by the new Government.

In the wake of the 1971 war, about 93,000 Pakistanis became prisoners of war in India. The Indian Army moved the POWs from East Pakistan to different locations in India. To keep the POWs informed of the welfare of their families in Pakistan, Radio Pakistan started a 12-minute programme 'Dua Salam' from 26 March 1972. This programme continued till the return of the POWs. As the messages increased, the programme duration was gradually extended to 45 minutes daily from 21 July. These messages were recorded at Peshawar, Lahore, Multan, Hyderabad, Quetta, Karachi and Muzaffarabad Stations. In addition, Radio teams visited many hometowns to record messages of families of POWs. During the first year Radio Pakistan broadcast about 31,000 messages in over 240 hours.

On the subject of POWs, Radio Pakistan also arranged broadcast of messages by well-known local personalities to their counterparts in India under the title 'Mera Paigham Mohabbat Hai Jahan Tak Puhnche' (My Message is Love for All). Some of the personalities who broadcast messages were: Josh Malihabadi, Abul Asar Hafeez Jallandhri, Faiz Ahmad Faiz, Sufi Ghulam Mustafa Tabassum, Ahmad Nadeem Qasmi, Mumtaz Mufti, Majnun Gorakhpuri, Akhtar Husain Raipuri, and Roshanara Begum.

From September 1973, PBC broadcast the names of the prisoners of war as they were repatriated to Pakistan. A running commentary, in a special national hook-up was broadcast direct from the Wagah border on the occasion of the arrival of the first batch of POWs.

ISLAMIC SUMMIT CONFERENCE 1974

Extensive publicity and coverage of the Islamic Summit Conference held in Lahore on 22-24 February, 1974 was arranged by all stations of PBC. The coverage was devised under four phases, i.e. introductory, pre-summit, the summit and post-summit. The pre-summit programmes stressed the significance of the Conference with particular reference to its political, economic and social aspects. The philosophy of Muslim thinkers, statements of national and local leaders, appropriate extracts from the national and international press and music of the participating countries were also broadcast. Curtain-raiser programmes on participating countries as well as on Palestine and the city of Lahore were broadcast in the national hook-up as well as in the External and World Services. During the Summit all stations of PBC remained on the air without break from 0630 hours to midnight from 21 to 25 February 1974. Six special bulletins in Arabic, French and English were broadcast daily for the duration of the Summit Conference.

The programmes broadcast in connection with the coverage and publicity of the Islamic Summit included: direct relay of over 45 hours of the opening and closing sessions via communication satellite on three circuits in Arabic, French and English; running commentary on the arrival of delegates in Pakistan from Lahore Airport, arrival of Heads of State/Government of Muslim countries for Juma Prayers at Badshahi Mosque and on the civic reception in their honour by the Chief Minister of the Punjab and a special two and a half hour Arabic Service daily from 21 to 25 February for the benefit of listeners in the Middle East. During this period, PBC broadcast a total of 2790 hours of programmes including interviews of 32 heads of delegation. It supplied recorded material of 3150 minutes to 35 countries containing Pakistani music, features and programmes on the Pakistani way of life. It also provided voice-cast facilities to 19 foreign broadcasting representatives for 66 hours. As broadcasting organizations in the Islamic world were depending on Radio Pakistan for prompt and detailed reporting of the Islamic Summit, the External Services put out almost round the clock broadcasts of its proceedings with exclusive interviews of leaders of delegations and other visiting dignitaries. Technical facilities were provided to representatives of Saudi Arabia to broadcast running commentary direct from the Airport on the arrival of King Faisal and on the Juma Prayers at Badshahi Mosque. Tape recorders and tapes were provided to foreign radio representatives. Sets of 10 LP records in presentation boxes, were presented to the Heads of Broadcasting Organizations of the participating countries.

QUAID-I-AZAM MOHAMMAD ALI JINNAH'S BIRTH CENTENARY

Quaid-i-Azam Mohammad Ali Jinnah's birth centenary was celebrated during 1976 and PBC organized special year-long programming. The significant national hook-up programmes broadcast in this connection included personal sketches of the life of the Quaid and interviews with his associates and contemporaries in the struggle for Pakistan.

In the national hook-up special programmes were aired on the inauguration of a mobile exhibition of the personal belongings of the Quaid, the International Congress on Quaid-i-Azam, foundation-stone laying ceremony of Quaid-i-Azam Museum and Centenary Memorial, special reference session of Parliament held to celebrate the Birth Centenary, etc.

Radio reports covering outside activities held by cultural and social organizations throughout the year were broadcast from all Stations. External Services also carried special programmes.

The material sent to foreign broadcasting organizations included a 30-minute programme entitled 'Meet Mr Mohammad Ali Jinnah' prepared in six languages—Arabic, Afghan-Persian, English, French, Persian and Turkish. The programme included speeches of the Quaid-i-Azam, national songs, folk and instrumental music.

ALLAMA IQBAL'S BIRTH CENTENARY

The birth centenary of Allama Muhammad Iqbal was observed in 1977. At the start of the year a radio report was featured on the inauguration of the centenary celebrations at the Allama's mausoleum in Lahore. The centenary celebrations programming included poetry readings and musical compositions of Iqbal's poetry in the national and regional languages and in the External Services.

Highlights of other programmes included discussions entitled 'Fikr-i-Iqbal ki Rahen' (Directions of Iqbal's Thought), Iqbal Memorial Lectures, quiz programme 'Jahan-i-Taza', series of talks entitled 'Dana-i-Raz' (One who knows Mysteries of Universe), Iqbal ka Ek Sher (A Verse from Iqbal), Iqbal aur Quaid-i-Azam, Iqbal Ek Ahad Afreen Sha'ir (Iqbal, an Epoch making Poet), Iqbal aur Bairuni Ma'asreen (Iqbal and Foreign Contemporaries), 'Iqbal-the Voice of the East', symposium on 'Reconstruction of Religious Thought in Islam', week-long programme (3-9 November 1977) recapturing various aspects of Allama Iqbal's life and works, and a programme under the title 'Iqbal Manzil Sialkot se Javed Manzil Lahore Tak' which covered all significant events connected with the life of Allama Iqbal.

700 ANNIVERSARY OF AMIR KHUSRAU

In connection with the 700t anniversary of the renowned poet, musician and sufi, Hazrat Amir Khusrau special programmes including series of talks, panel discussions, musical features, music concerts, symposia, *musha'aras* and plays were broadcast. Radio coverage of the celebrations in the country was broadcast from all stations of Pakistan

Broadcasting Corporation, both in the national and regional languages. Extensive coverage was also given to the International Congress on Amir Khusrau organized during the period.

1400 YEARS OF HIJRA

The end of 1400 years of *Hijra* and ushering in of the fifteenth century was celebrated in a befitting manner during the last year of the fourteenth century. It was also observed as the year of Al-Quds al-Sharif. Radio Pakistan covered the celebrations and broadcast special programmes on Islam and social development and uplift.

Programmes highlighted major events in Islamic history, the contribution of Muslim scientists, philosophers, mathematicians, military strategists and other pioneers in every field of life towards the evolution of civilization.

RADIO AWARDS, 1982

In 1982, Radio Pakistan awarded prizes for the first time to those outstanding artists, writers and producers who had made remarkable contribution in radio programmes during the past five years and who had not received recognition. The following persons were given awards at a ceremony held in March 1982 in Rawalpindi:

S.	Category	Recipient.
1.	Qari	Obaidur Rahman
2.	Naat Khwan	Qari Waheed Zafar Qasmi
3.	Religious talks	Shah Balighuddin
4.	Playwright	Bano Qudsia
5.	Playwright Regional	Ata Shad
6.	Drama Artist Male	Mahmood Ali
7.	Drama Artist Female	Sajida Syed
8.	Drama Artist Regional Male	Irfan Khusat
9.	Drama Artist Regional Female	Mehnaz Begum
10.	Stock Character National	Amir Khan
11.	Stock Character Regional	Abdullah Jan Maghmoom
12.	Announcer/Compere National	Nilofer Abbasi
13.	Announcer/Compere Regional	Nighat Iqbal

14. Newsreader English	Shaista Zaid
15. Newsreader Urdu	Khalid Hameed
16. Newsreader Regional	Kamran Bhatti
17. News Reporter	Tanwir Siddiqui
18. Sports Commentator	Iftikhar Ahmad
19. External Service Broadcaster	Hla Maung
20. Classical Music	Hameed Ali Khan, Fateh Ali Khan
21. Qawwali	Nusrat Fateh Ali, Mujahaid Mubarak Ali
22. Light Music	Mehnaz
23. Folk Music	Sain Akhtar Husain
24. Producer, Religious Programmes	Abdul Hayee Qureshi
25. Producer, Current Affairs	Hasan Zaki Kazmi
26. Producer, Drama/Features	Muhammad Anwar Cheema
27. Producer, Miscellaneous	Azim Sarwar
28. Producer, Radio Report	Fakhr-i-Alam Naumani
29. Producer, Music	Inam Siddiqi
30. Producer, Farm Forum	Sikandar Baloch
31. Recording Engineer	Muhammad Rafique
32. Best Engineer	Riazul Haq

FOURTH SAF GAMES

The fourth South Asian Federation (SAF) Games were held in Islamabad from 20 to 27 October, 1989. Sportsmen from all the seven member countries of the SAARC participated in these Games. PBC covered the event for the benefit of listeners abroad. Informative programmes about participating countries, games and athletes were started from 9 October in the national hook-up and locally from each station. PBC set up its own studio, control room and commentary box at Jinnah Stadium, Islamabad and broadcast running commentaries and special radio reports.

A three-hour live coverage of the opening ceremony performed by the President of Pakistan was broadcast. It was followed by the special sports transmission which was on air from 9 a.m. to 11 p.m. daily. During the eight days of the Games, 112 hours of programmes were broadcast in the national hook-up. About forty radio reports were aired daily ending with the final results of each event. On the last day twelve round-ups on different games were broadcast including interviews of gold medallists and final results. The closing ceremony received live coverage. Radio reports and running commentaries on other events and venues were also presented from the Liaquat Gymnasium, Practice Hall Islamabad, Army Ground Rawalpindi, NAFDEC Cinema Hall Islamabad, Wah Forum and Army Squash Centre, Peshawar.

SEVENTH WORLD CUP HOCKEY TOURNAMENT

The Seventh World Cup Hockey Tournament was held at Lahore from 12 to 23 February, 1990. A World Cup cell was set up at Lahore Station of PBC and introductory programmes were broadcast before the Tournament. Live coverage of all the 42 matches played in the Tournament was broadcast, with a summary of the day's matches given daily in the national hook-up programme '*Alami* Sports Round-up'. Special interesting and informative programmes on the background and history of the game were shown during the intervals between games. AIR team was provided necessary facilities for coverage. A BBC representative was also present who sent periodic reports direct to London.

SAARC QUIZ, 1990

Pakistan's offer to hold the first SAARC Quiz was accepted and PBC was entrusted with the organization of the event. The Corporation in collaboration with the Ministry of Foreign Affairs organized the show from 25 to 27 February, 1990. According to the SAVE Committee, each participating country had to select a team of two University students in three stages i.e. divisional, provincial and national. The two-person winning team at the national level was to compete in the SAARC Quiz. Besides the selection of the team, PBC undertook other tasks, most important of which was composing quiz questions, and judge selection.

The Quiz itself was held in three stages. Draws were held to form teams for the Quiz on 25 February 1990. In each team the two members had to be from different countries. The recording of the Quiz was broadcast by all member countries after permission of the SAVE

Committee. All the seven member countries (Bangladesh, Bharat, Bhutan, Maldives, Nepal, Pakistan and Sri Lanka) took part in the Quiz.

Golden Jubilee of Pakistan Resolution

There is no recording available of the historic public meeting at Lahore on 23 March 1940 where the Lahore Resolution (commonly known as Pakistan Resolution) was passed demanding an independent state for Muslims of the subcontinent. Radio Pakistan re-enacted a 'sound picture' that was first broadcast in the national hook-up on 23 March 1981. It was re-broadcast in 1990.

The year 1990 was declared the Golden Jubilee Year of the Lahore Resolution, and PBC organized elaborate programmes during the whole year. These included the usual discussion and documentaries, *musha'aras*, musical features, quizzes, interviews of veteran leaders present at the historical meeting, and workers of the Pakistan Movement from all walks of life. Competitions were held at district, divisional, provincial and national levels, with winners of each stage going on to the next stage. These included patriotic songs, *naat* and *Qirat* competitions, speech contests and quiz programmes. The finals of these competitions were held at Islamabad. The final of the Independence Quiz was held in Islamabad on 12 December 1990 between the winning teams at Provincial level from the four Provinces of Punjab, Sindh, NWFP and Balochistan as well as from Azad Kashmir, Northern Areas and the Federal Capital.

Independence Day Celebrations

From the late seventies, the Independence Day celebrations have been large-scale events spread over two weeks beginning the first of August. All PBC Stations broadcast special programmes during the period, highlighting the struggle and contribution of leaders and workers of the Pakistan Movement. Extensive coverage is given to functions held on the occasion around the country.

Stations arrange competitions particularly for students and the youth, and award prizes in various categories including the highly publicized prize for best illuminated building in the cities. A day-long stage show is arranged at the main gate of each Broadcasting House for the general public.

ELECTION COVERAGE

Elections are an important democratic tradition and have a crucial impact on the destiny of a nation. Radio Pakistan makes special arrangements to cover an election in the country. An election cell is set up at the radio headquarters headed by the Director of Programmes under the direct supervision of the Director General. The Cell issues detailed instructions for election coverage to all stations and units. Each station is required to set up an election cell comprising the Station Director as its head, with a Programme Manager, News Editor and Senior Producer as its members. This cell is responsible for all matters connected with elections. The coverage is normally divided into six phases. The first phase relates to the publicity of the election schedule, i.e. candidate nomination, withdrawal of candidature, allotment of election symbols, appeals and dates of polling.

The second phase is pre-election publicity. During this phase the benefits of the democratic system are highlighted. People are asked to vote for persons of good character with a background of selfless service to the nation. This is done through talks, discussions, and interviews with prospective candidates.

The third phase offers training for voters and this continues till the election day. During this phase the election process is explained in detail from the rights and obligations of candidates, information about constituencies and polling stations, to the importance of a peaceful election. Election commercials encourage voters to exercise their right to vote.

The fourth stage is on the day of the election. Stations begin coverage from early morning till the start of polling, interspersed with popular music, *naats*, and patriotic songs. Slogans are broadcast urging people to vote for the best candidate from the national point of view. When polling starts all Stations join a special national hook-up that continues till the polls close. PBC representatives are posted at about 100 polling stations throughout the country giving live reports on the election. For Provincial Assemblies a similar election programme is presented by Radio Stations in the provincial capitals.

The fifth stage relates to coverage of election results, which begins at 9 p.m. and continues till final results are in. A booth is set up in the Election Commission so that results are broadcast the moment they become available. Provincial Assembly election results are also announced from Islamabad in four different Provincial hook-ups. The results are announced in the national and regional news bulletins, World Service and the External Services.

The sixth and last phase is the post-election programme, wherein people are advised to cooperate with the elected representatives to work for the good of the nation.

Prior to the coverage of the March 1977 Election campaign, the Lahore High Court and the Supreme Court passed an order directing PBC and PTVC to give equal coverage to the election campaign of both major contestants in the elections, i.e. the Pakistan Peoples Party and the Pakistan National Alliance. The PNA must be allowed to present their manifesto and address the nation through the media on an equal footing with the PPP.

The High Court also directed that both contenders in the election be given equal coverage. The only exception would be the Prime Minister's address to the nation in his capacity as leader of the government speaking on important national issues, as opposed to his position as Chairman PPP, electioneering for his party.

The Supreme Court on a petition from PBC and PTVC partly modified the order of the Lahore High Court substituting it with their order for the two organizations to follow in the discharge of their functions till the hearing of the main petition. Among other things, the Supreme Court order stated:

> While due to diversity of circumstances, such as availability of news at a given time or presentability of the news on the national broadcasting and television system, it may not be possible to allocate equal time to the political parties including the Pakistan Peoples Party and Pakistan National Alliance, group of persons or individuals seeking election, an attempt should always be made to keep a balance in broadcasting and telecasting their activities with regard to the election campaign in their daily news bulletins and other releases including live voice of political leaders addressing public as far as possible.
>
> In our opinion a compliance with the general directions issued to the petitioners in the above terms will enable them to function in letter and spirit, within their own charter without in any way placing any restrictions on the direction vested in them under the law.

The elections of 3 February 1997 marked the first time that Federal and Provincial Assembly elections were held on the same day. Radio coverage was similar to previous elections, featuring special programmes including: interviews of the leaders of major political parties in the national hook-up under the title 'Siyasi Jamaaten aur Manshoor' (Political Parties and Manifestos), series of discussion programmes entitled 'Election ke Taqaze' (Requirements of Elections), six programmes on 'Elections aur Hamare Masail' (Elections and our Problems) where political experts answered phone-in queries from listeners.

Actuality reports sent by PBC representatives from 81 points in the country were broadcast in the national hook-up (85 reports) from 9 a.m. to 5 p.m. and an equal number of reports were put out from the regional stations. The second national hook-up transmission started at 8.30 p.m. for the announcement of results and continued for about 38 hours. Election results were received in the studios at the 'Radio Election City' in PBC Islamabad, which was fitted with fax, computer and telephone facilities. From 11.30 p.m. onwards a special bulletin giving latest results and party positions was broadcast every half hour till the final results were announced.

Floods

When natural disasters occur such as the floods which devastate the Provinces of the Punjab and Sindh, all PBC stations are put on alert, with 24-hour flood reporting desks set up, special teams deputed to the flood-affected areas for on the spot coverage and transmission hours are extended. Round the clock transmission is arranged for several days to keep alive what is very often the only communication link left with affected areas.

The staff never fails to rise to the occasion. Special programmes are broadcast to provide solace and psychological support to people affected by the floods and to raise public morale with up to date radio reports, eye-witness accounts, special news bulletins and weather reports. Professional advice to the affectees was provided to help them deal with their hardship and plan for the future. Constant liaison was maintained with the relevant departments of the Provincial Governments and Army authorities responsible for flood rescue and relief operations as well as rehabilitation work.

In the aftermath of the 1973 floods, Radio Pakistan altered its regular programming and broadcast schedules to suit the need of the crisis. Due to a breakdown of the telecommunication system, transmitters had to be operated. Even the news agencies used the facility of PBC transmitters during the first two days of the breakdown. An emergency flood control cell was set up at the Headquarters of PBC and all the regional units and officers were put on duty round the clock. Special programming included news updates of the flood situation, road and rail conditions, SOS messages from displaced persons, and information on prevention of diseases and epidemics. The Central Productions arranged special performances by singers at no charge. Fourteen commercial LP records were recorded through EMI with the sale proceeds donated to the relief fund.

An identical action plan was taken during the floods of September 1988 when a ten foot high wave hit a number of villages in the suburbs of Lahore. On the night of 26 September 1988, the Flood Control Centre issued an emergency flood warning that was announced by the Station in its programmes and news broadcasts. The magnitude of the flood devastation was such that Lahore Station suspended regular broadcasts for a week and devoted its entire transmission to coverage of the flood situation. As usual PBC was the only communication link between the flood victims and the Provincial authorities. A flood cell was set up at the Lahore Radio Station. Radio representatives were sent to various locations from where they sent situation reports.

Broadcast content included updates on the flood situation, announcements and interviews of relevant authorities about relief camps, messages from social welfare organizations, appeals for blood donation, announcements about road and rail traffic, reports received from the Deputy Commissioners of Sheikhupura, Kasur, Sialkot and Sahiwal, relief operations of government and agencies, expert advice on preventive action regarding crops and livestock. Interviews of the Governor, Chief Minister, Federal and Provincial Ministers were also broadcast.

Messages from relatives of missing persons were put on the air, along with responses. On 12 October, Lahore Station organized a special benefit Radio Cricket Show involving the entire Pakistan Cricket Team, top ranking artists and cricket fans. Prizes and proceeds were donated to the relief fund.

As the flood waters reached Bahawalpur and Bahawalnagar area through the Sutlej River, emergency measures were taken to meet the grave situation. For its part, PBC's teams were sent to all the three Headworks in the Division. The first requirement was the evacuation of people from the bed of the river. After consultation with the local authorities frequent announcements were made asking people to move to safer places. This timely action resulted in the saving of thousands of lives and property.

Bahawalpur Station normally broadcasts only in the evenings but it began a morning transmission from 9 October. Multan Station and the national hook-up programme 'Subh-i-Pakistan' which has a very wide listening aired special announcements. Reports on the latest flood situation from different Headworks and other strategic places were broadcast after every 10-15 minutes. The situation became so critical that the Station continued its transmission even after the usual closing time of 11 p.m. Latest reports from the Divisional Flood Cell, Army Flood Cell and from Radio representatives about discharge of water, road and railway conditions were broadcast intermittently.

When the emergency situation eased, attention was directed to relief operations, and these broadcasts spurred philanthropists to send trolleys full of relief goods including clothing, foodstuff, wheat, medicines, etc. to the Radio Station for dispatch to relief camps. Bahawalpur Station also broadcast programmes for the benefit of areas threatened by the floods. These included Ahmadpur East, Vehari District, and Jalalpur Pirwala.

Again the same plan of action was employed in the September 1992 floods. Multan and Bahawalpur Stations were located in the flood-affected areas. This time advance warnings and information were available so that there was comparatively less loss of life, property and livestock. The floods hit populated areas on 10 September 1992 and on the night between 14 and 15 September the transmission was extended and continued without break to give updates on the flood situation and other related information. Radio representatives risked their lives in flooded areas to send in reports from roof tops surrounded by 6 feet of water. They cooperated with Army authorities and local Administration. Hyderabad Station set up a temporary studio *Mehran Mauj*' on the bank of the River Indus near Kotri Barrage. This gives an idea of the involvement of Radio Pakistan in national affairs, and its aim to be with the people in good times and hard times.

During natural calamities like floods and earthquakes, power cuts are a common feature even in advanced countries. The Radio and TV stations have their own power generating plants but due to the power failures in affected areas, victims cannot access television. In such a situation radio, the common medium of information is accessible through transistors. This important feature of Radio accessibility needs to be universally acknowledged. A point in example is the Great Hanshin Earthquake in Kobe, Japan on 17 January 1995 when power supply in the city was cut instantaneously. When people left their homes for safety areas they took their transistors with them.

Railway Accidents

Pakistan Railways has had some very serious railway accidents for example those at Jhimpir, Sangi (3-4.1.1990), and Ghotki (7-8.6.1991).

In such events the nearest Radio Station organizes three or four teams for coverage immediately on receipt of information of the accident. One team goes to the site of the accident for on the spot recording. Another is deputed to the hospitals where the victims are taken. The third team contacts relevant agencies and authorities involved in rescue and relief operations, e.g. railway authorities, local administration and welfare organizations. Yet another team is located at the Radio Station to provide updates and information and man emergency telephone lines.

Names of victims are announced in the national hook-up, as well as radio reports of relief and rescue work. This special reporting continues for two or three days after the accident. The local Radio Station gives reports after short intervals and makes appeals for blood donation.

Karachi Tragedy

On 14 July 1987 two massive bomb blasts in Karachi caused heavy casualties. The news of the bomb blasts reached the Broadcasting House a few minutes before 7 p.m. Regular programming was immediately changed and film music and light entertainment was replaced with sombre programmes. Calls for surgeons to report on duty and for blood donation were broadcast frequently. Two teams of Producers were sent out to the scene of the tragedy and to hospitals to prepare situation reports. A detailed 8-minute bulletin about the tragedy was broadcast at 11.08 p.m. immediately followed by the first situation report comprising interviews with victims, eyewitnesses, doctors, relief workers, etc.

The following day's programming was also sober. All light music was replaced with *naats*, *qawwalis*, selected *kalam-i-Iqbal* and morale-boosting patriotic songs. Nine situation reports were broadcast during the day. Appeals and announcements about blood donations, traffic conditions, cooperation in relief work, arrangements for identification of bodies, etc. were made throughout the day. A special 30-minute programme '27 hours since the Karachi Tragedy' was broadcast in the national hook-up after the 9 p.m. news bulletin and commentary. This included the details of the President's visit to the hospital and affected area, latest situation report and details of relief work being carried out. Special programmes were also broadcast about the need for unity, faith and discipline, morale building and determination to fight the menace of terrorism at all levels. Coverage was given in the news bulletins and in regional languages.

OJHRI CAMP EXPLOSION

Tragedy struck in the Islamabad/Rawalpindi area when an accidental explosion occurred in an ammunition depot on the morning of 10 April 1988. Radio Pakistan responded to this unforeseen catastrophe promptly with an emergency bulletin broadcast less than an hour after the incident. The nation received the news that an accidental fire had broken out in an ammunition depot in Rawalpindi. News reports followed the developments in the story and this helped in curbing unfounded rumours of terrorist involvement and allayed apprehensions in the minds of the people.

Inspiring programmes were broadcast to emphasize the Islamic traditions of fortitude and restraint in times of trouble and widespread suffering. PBC also broadcast messages of people searching for missing loved ones. At the same time detailed information was given about the relief and rescue operations carried out by the local authorities. Radio's role in this crisis was highly appreciated, and particularly lauded by the President of Pakistan, who in a letter dated 15 April 1988 to the Director General of PBC, said:

> It is a source of personal satisfaction to me that your organization responded to this unforeseen tragedy with admirable promptitude and sense of

dedication.... I would like to convey my personal appreciation to you and all those who were engaged in this noble campaign round the clock with exemplary dedication and diligence during the emergency without caring for their personal safety and comfort. I trust Radio Pakistan will continue to maintain its high standards as a vital medium and keep on serving the people of Pakistan and our countrymen living abroad through its internal and external services consistent with its excellent traditions extending over many years.

DEATH OF PRESIDENT ZIA

President General M. Ziaul Haq died in an air crash near Bahawalpur in the afternoon of 17 August 1988. Radio Pakistan was the first to broadcast this tragic news in its main bulletin at 8 p.m. from where it was picked up by news agencies and broadcasting organizations. The bulletin gave a life sketch of the deceased President, his achievements particularly with reference to establishing an Islamic identity for the nation, such as the Shariat Bill, independence of Afghanistan, working towards Islamic unity, and establishing peaceful and cordial relations with neighbouring countries. It also gave the names of all the persons who were in the ill-fated aircraft.

All stations remained linked in a special national hook-up that contained *tilawat*, selected *hamds* and *naats*, extracts from his speeches, recitation of Allama Iqbal's poem '*Masjid-i-Qurtaba*' in the voice of Syed Zulfiqar Ali Bokhari. This special programme session continued till the address to the nation of the Acting President. From the next day special programmes were broadcast in the national hook-up as well as locally in the national and regional languages. These programmes were prepared and broadcast under the title 'Tab-o-Taab-i-Javedana' (Everlasting Glory) containing details about the achievements of the late President, his multi-dimensional personality as well as life sketches of all the persons who lost their life in the air crash.

All commercial programmes, advertisements and sports items were stopped for the duration of the special broadcast. In addition, condolence messages from abroad, arrival of foreign heads of state and government and other dignitaries and delegations, information about the funeral and burial were also broadcast. Radio Pakistan broadcast a four hour running commentary on the funeral procession and burial of the President. It is considered to be the longest commentary in the history of broadcasting for such an occasion. Similar programmes were carried by the World Service and the External Services particularly the Arabic, Dari, Irani, Turkish and French Services. This sober programming continued up to the end of the 10-day official mourning period on 26 August.

It may be of interest to know that when the news of the President's death reached Tokyo, Radio Japan had closed its transmission. Radio Japan re-started its transmission to announce the tragic news in a solemn broadcast. It was for the first time in the history of Radio Japan that a transmission was reopened after it had signed off.

Afghan Refugees

When the USSR invaded Afghanistan causing the influx of three million Afghanis into Pakistan, Radio Pakistan began a special session of reports and programmes that evoked a spirit of Islamic brotherhood and empathy among the people of Pakistan towards the unfortunate refugees. These refugees were exhorted to accept whatever help was available and to adjust to their new circumstances with fortitude. The daily programme 'Hindara' of Peshawar Station was used to convey these messages, in addition to a number of programmes started from Peshawar and Quetta Stations, and in the External Services particularly the Dari and Hazargi Services.

Bhal Safai (Dredging)

Pakistan has one of the most extensive canal systems in the world. It consists of 14 barrages and about 23000 miles of canals—large and small. Removal of silt from the canals is an annual feature, but in 1992, a thorough cleaning of the system was carried out. The campaign continued for the entire month of January. The clean-up was inaugurated by the Prime Minister of Pakistan and was overseen by the Chief Minister of the Punjab.

Radio Pakistan Lahore cooperated with the Provincial Government to publicize the campaign. Special thematic songs were produced and presented on the air. Radio representatives visited sites along the clean-up route in the Province and sent their reports for broadcast, capturing the atmosphere and detail of the work in progress. Lahore Station broadcast interviews of farmers, agriculturists, public representatives, workers and officers of concerned departments involved in the campaign. The Chief Minister was so pleased with the performance of Radio Pakistan that in token of appreciation he presented a van to Lahore Station.

In 2000, the Army personnel got involved in the canal clean-up programme with the result that more canals were cleared in a systematic manner at much less cost. Radio Pakistan provided information on the progress of the work as well as views and advice of agricultural experts. The farmers especially those living on the outskirts of the canal network were urged to actively participate in the clean-up campaign.

ISLAMIC SUMMIT 1997

An extraordinary session of the Islamic Summit was held in Islamabad on 23 March 1997. Radio Pakistan's plan to cover the event consisted of three phases. The first phase starting on March 10 had introductory programmes about the member countries entitled 'Ek Hon Muslim' (Muslim World Unite), with music features on the national hook-up. It also included eyewitness accounts on the arrival of all the foreign Heads of State and delegations. This commentary was broadcast direct from the Islamabad International Airport from 8.30 a.m. on 22 March 1997 to 1.30 a.m. the next day.

The second phase comprised direct relay of the proceedings of the opening and closing sessions of the Summit Conference; an all-Pakistan *musha'ara*, interviews of heads of delegations, special news bulletins, extensive coverage in the World and External Services as well as in the daily current affairs programmes. The third phase focussed on the joint declaration, outcome of the Conference, economic and social achievements, interviews of heads of delegations, etc.

PBC set up a special studio in the Convention Centre Islamabad, in cooperation with the Civil Aviation Authority. A media centre was also set up in collaboration with PID to provide voice-cast, transmission, recording, satellite circuit, telephone and fax facilities to media personnel from member countries.

Golden Jubilee Celebrations

The Golden Jubilee of the birth of Pakistan was celebrated throughout the year 1997. PBC broadcast special programmes on the occasion which included discussions, interviews, seminars, *musha'aras*, music concerts, radio shows, drama festivals, stage shows, thematic songs, special programmes for the youth, women and children, quiz and elocution contests, talks and features and listeners views.

PBC gave live coverage of the flag-hoisting ceremony and the Prime Minister's address to the nation held in front of the Parliament House on 14 August 1997. Extensive coverage was given to all national and local functions held in the country in connection with the golden jubilee celebrations. A Pakistan flag measuring 170 ft x 110 ft made by Pakistan Air Force was unfurled at the PBC Headquarters building on 13 August 1997. It was considered to be one of the biggest Pakistani flags in the world.

Golden Jubilee of RP/PBC

The Golden Jubilee celebrations of Radio Pakistan were held from 19 May 1998 to 27 June 1998, and were inaugurated by the Prime Minister. A seminar 'Radio Pakistan *Ehd ba Ehd – Ek Jaiza*' (Review of Radio Pakistan's Performance) was held the same day and was presided over by the Minister for Religious Affairs, with participants including the Information Minister and former Directors General of Radio Pakistan/Pakistan Broadcasting Corporation. A feature under the same title was also presented. All the Directors General received awards in recognition of their services to the cause of broadcasting. The President of Pakistan chaired a seminar on the 'Role of Radio Pakistan in National Identification and Integration' held on 21 May 1997 in Lahore.

Among the special programmes broadcast on the occasion were seminars on the contribution of Radio in the fields of Arts and Literature and national development. The last function was held at the Convention Centre Islamabad on 27 June 1998. The all-Pakistan *musha'ara* broadcast on 22 March 1998 was in two parts, covering poetry of living and deceased poets. Special programmes were broadcast in regional languages as well as in the World and External

Services. The monthly *'Ahang'* published a special combined issue for August-September 1998 to mark the celebrations.

CENTENARY OF THE SAUDI KINGDOM

A two-day (27-28 January 1999) International Conference was held in Islamabad on 'Saudi Arabia as a Source of Strength and Stability in the Muslim World' as part of the centenary celebrations of the Kingdom of Saudi Arabia. Special programmes were broadcast including exclusive panel interview of the Ambassador of Saudi Arabia, discussion on *'Islami Dunya ki Falah-o-Behbood ke liye Saudi Arab ka Kirdar'*, feature in World Service, radio reports on sessions of the Conference and talks in regional languages from different Stations.

YAUM-I-TAKBEER

PBC broadcast special programmes in connection with the observance of *Yaum-i-Takbeer* from 11 to 28 May 1999, highlighting Chaghi as a symbol of success, self-reliance and advancement in science and technology. Also stressed was Pakistan's ability to overcome economic difficulties created by unjust sanctions imposed by some international institutions and certain countries. Special programmes covered subjects such as Atomic Capability—A Guarantee for Peace and Security, Advancement in Science and Technology, Defence and Economic Self-Sufficiency and Balance of Power in South Asia.

The morning network programme 'Subh-i-Pakistan' revived the historical and memorable happenings during the same period last year, impressions of cross-section of people under the title 'Faulad Hai Momin' (Sturdy is the Believer) in the national and regional languages, special songs/jingles in the form of children's choir and radio reports daily from regional stations were organized. Every effort was made to provide maximum coverage to events organized by various Federal Departments/Divisions, Provincial Governments including Azad Kashmir Government and Northern Area Administration. The programmes were aired in the national and all the regional languages. 28 May 2000 was observed as Science and Technology Day.

Census

The work on the 5th Population and Housing Census was carried out in the country from 2 to 18 March 1998 although it was due in 1981. Radio Pakistan gave it full publicity in its programmes and commercials. Special programmes were arranged in the national and regional languages to inform the public on the various aspects of the Census held after 17 years. Stations broadcast locally quiz competitions, besides commentaries and discussions highlighting the importance of the Census, to create awareness among the masses about its significance in socio-economic development, and motivate them to provide accurate information to the census workers. Exclusive interviews of the Chief Election Commissioner and the Election Commissioners were also broadcast.

Bicentenary of Tipu Sultan

The 200th death anniversary of the great freedom fighter, Tipu Sultan *Shaheed*, was observed with solemnity and national enthusiasm. PBC organized special programmes in the national as well as regional languages highlighting the life and achievements of the great patriot. The programmes included a play 'Shaheed-i-Seringapatam' and a seminar entitled 'Tipu Sultan aur Islami Falahi Riyasat' (Tipu Sultan and Islamic Welfare State).

Golden Jubilee Celebrations China

As part of the week-long (18-25 September 1999) national programme to mark the Golden Jubilee of the People's Republic of China, PBC aired a number of programmes on its network, including a two-round quiz competition on Pak-China friendship for University students at major stations of PBC. The winners of these competitions took part in the final quiz competition held at PBC Islamabad. A documentary on Pak-China relations over the past 50 years was aired and live coverage of the function held at Convention Centre, Islamabad that was chaired by the President of Pakistan. All stations including FM 101 accorded appropriate coverage to the cultural activities of the week-long celebrations.

PBC Awards, 1998-99

PBC presented awards on national and regional levels for outstanding performances. Functions were organized at the Federal and the four Provincial Capitals and awards were presented in various categories. A Super Star Award for excellence in performance during the past half a century was presented, and only living nominees were eligible. Other awards presented were an Excellence Award on the basis of performance during 1998-99, Lifetime Achievement Award for outstanding/legendary role in a field of broadcasting, and Special Award to persons having excellent performance record.

A fourteen-member jury comprising renowned intellectuals, poets, writers, educationists and media experts was appointed for the selection of winners of these awards. The jury adjudged the performance and quality of some categories such as the writers, producers, news editors, reporters, announcers and engineers. Among the Excellence Award winners, artists in twelve categories such as *naat khwani*, newsreading, singing, drama acting, compering and sports commentary, etc. were chosen through listeners' polls.

On the national level as many as thirty-four Super Star Awards were presented to celebrated artists of Radio Pakistan in categories such as religious programmes, music, drama and other fields. Seven Special Awards went to writers, producers and other professionals associated with Radio Pakistan in recognition of meritorious services rendered in the broadcasting field. Three persons were decorated with the Lifetime Achievement Award. Sixty one writers, performers, qaris etc. including 12 selected by the listeners, received Excellence Awards. The total number of recipients of Awards on the national level comes to 105, a complete list of which is given in Appendix III.

The function to honour artists and producers on the national level was held at Islamabad on 4 December 1999. Persons who could not attend the function at Islamabad received their awards at the regional functions. The regional functions were held at Peshawar, Lahore, Karachi and Quetta on 12 December 1999, 31 January 2000, 20 February 2000 and 11 March 2000 respectively.

Millennium Programmes

PBC put across special programmes to mark the end of the second millennium and to welcome the new millennium. Under the series 'Ek Hazar Saal—Sadi ba Sadi' (One Thousand Years—Century by Century) ten documented features were aired with actualities in the form of interviews, extracts, opinions of intellectuals who reviewed the important events including ideologies, trials, personalities, discoveries and inventions occurring in a particular century starting from the beginning of the second millennium. Each programme covered one century and carried in-depth documented study starting from the first year to the end of that particular century. Ten programmes were put on air under the caption 'Beesween Sadi—Ashra ba Ashra' (Twentieth Century—Decade by Decade), giving details of important events in each decade of the 20th century. Other special programmes broadcast included: 'Sur Samandar' dealing with development in music during a century in musical form, Radio Millennium Lectures, Shahkar (Hit) Plays, Phone-in Programmes and special programmes in the morning network disc jockey programme 'Subh-i-Pakistan'.

Convention on Human Rights

Radio Pakistan broadcast live the proceedings of the Convention on Human Rights and Human Dignity including the address of the Chief Executive on 21 April 2000. Human rights was a main theme in special programmes organized, including thematic plays. These programmes were broadcast in both national and regional languages from all stations. The points projected through these programmes included improvement in the quality of life of ordinary Pakistanis, elimination of bonded labour, ban on use of fetters for prisoners, treating 'honour killing' as murder, etc.

Drought in Sindh and Balochistan

Radio Pakistan broadcast special programmes during the drought in Sindh and Balochistan. PBC stations in these two provinces made every endeavour to keep the people abreast of the situation and the relief measures taken by the Government. The Chief Executive's appeal to donate generously to the Drought Relief Fund was broadcast and his visit to the affected areas was covered by Radio Pakistan's representatives who travelled in the helicopter with him.

Radio Pakistan also launched a seven-day campaign to raise funds for the drought-affected people of Sindh, Balochistan and Cholistan from 7 to 13 June 2000. All PBC stations including World Service and FM 101 aired eight-hour transmission starting at 4 p.m. and continuing till midnight everyday. Renowned artists and singers participated in music shows arranged by different stations. The External Services and the World Service also broadcast special programmes and appeals for donations. Similar fund-raising programmes were organized by the three FM stations also. For their part, PBC employees donated a day's wages for the relief fund and the cheque was presented by the Director General, PBC to the Chief Executive for the Drought Relief Fund on 9 August 2000.

EXCELLENCE AWARDS 1999-2000

The ceremony for the national Excellence Awards was held at the Liaquat Memorial Hall, Rawalpindi on 14 April 2001. As many as 47 Awards were presented to celebrated artists, writers, producers and other professionals associated with Radio Pakistan. Writer Ahmad Nadeem Qasmi and singer Mehdi Hasan were honoured with the Lifetime Achievement Award. Singer Tarannum Naz Laila was given the newly constituted Melody Queen Noor Jahan Award along with a cash prize of one lakh rupees. Multan Station was declared to be the best model station of PBC for the year. The list of recipients is at Appendix IV.

Earlier ceremonies for the regional Excellence Awards were held at Quetta (10 November 2000), Peshawar (18 November 2000), Karachi (06 January 2001) and Lahore (13 January 2001).

LOCAL BODIES ELECTIONS

The local bodies elections-2000 were held in phases, and PBC launched a campaign to inform people about the schedule of these polls. The points stressed during the campaign included preparation of a new voters' list and urging heads of families to register their family members, including females, who were eligible to vote. Also to inform people of the fundamental themes of the local government plan and of the advantages of the devolution of power, which aimed at empowering them to determine their own priorities for development at union council, *tehsil* and district levels. People were to be motivated to cast their votes in these elections enthusiastically. Special transmissions were arranged during polling.

OTHER ANNIVERSARIES

Some of the other important anniversaries observed by Radio Pakistan include: 25th Centenary of Lord Buddha, 500th Anniversary of Guru Nanak, 100 Years of Sindh Madrassatul Islam, 100th death anniversary of Mirza Ghalib, Silver Jubilee (*Jashn-i-Simin*) of Radio Pakistan (August 1972), 26th Anniversary of External Services (August 1975), Silver Jubilee of Central Productions (December, 1985), 200th birth anniversary of Mirza Ghalib (March 1999), and Silver Jubilee of Bahawalpur Station (August 2000).

12

STRUCTURE AND ORGANIZATION

POSITION AT INDEPENDENCE

At the time of Independence, Radio Pakistan was a non-profit State-controlled organization operating under the auspices of the Ministry of Interior (Information and Broadcasting Division) of the Government of Pakistan. The chief executive was called the Controller of Broadcasting.

The three small radio stations existing at that time functioned under very trying conditions. There was acute shortage of staff. About 473 staff members of AIR had opted for Pakistan, but many had difficulties reaching their destinations. The Headquarters, the regional stations and news units of Radio Pakistan were short of 33 officers and 71 non-gazetted staff. There were only 19 engineers against the need for 50. The staff had to work under extremely difficult conditions. There was no office or residential accommodation, no furniture and no office equipment. Seating space was inadequate.

The following are some of the officers who braved the spartan conditions prevalent in August 1947:

Controller of Broadcasting	Z.A. Bokhari
Director of Engineering	Riaz Ahmad
Deputy Controller of Broadcasting	Rashid Ahmad
Director of News	Mohamad Sarfraz
Deputy Director of Engineering	Bashir Ahmad
Station Director, Dhaka	G.K. Farid (Took charge as SD Lahore on 1.10.1947)
Station Director, Lahore	Ahmad Salman (Promoted as Deputy Controller of Broadcasting w.e.f. 7.10.1947
Station Director, Peshawar	S.S. Niazi

Public Relations Officer	Capt. Ahsanul Haque
Asst. Station Directors	N.M. Rashed Zainul Abedin
	M.S. Rahman Asnain Qutb
News Editors	Shaikh Ihsanul Haque
	A.G. Eirabie Hamid Jalal

The Directorate General was responsible for administration and problem-solving in the various constituent units of Radio Pakistan. It also served as an effective link between the Government and service units of Radio Pakistan. On 14 August 1949, the Central News Organization, responsible for collecting, editing and broadcasting news in the home and external services of Radio Pakistan was centralized at Karachi after a high power transmitter had been installed. The News Units at Dhaka, Lahore and Peshawar were set up immediately after Partition on 15 August 1947. After the centralization of CNO at Karachi, the News Unit at Lahore was wound up and a skeleton staff was left at Dhaka and Peshawar to cater to the local requirements. The News Unit at Lahore was re-established in October 1950. A News Unit was set up at Quetta in December 1951.

The budget for the year 1953-54 was drastically cut. To keep expenditure within the sanctioned budget grant several measures were taken including the decision to postpone some of the development schemes, to reduce allowances to artists by 40 per cent, and other perks like a travelling allowance, telephone and furniture, etc. Recruitment was curbed. It was decided to disband the News Unit at Quetta, the Staff Training School and the Listener Research Organization at Dhaka and to reorganize the regional news units.

Senior postings as on 14 August 1950 were:

Controller of Broadcasting	Z.A. Bokhari
Director of Engineering	Riaz Ahmad
Deputy C.B. (Prog)	Rashid Ahmad
Deputy C.B. (Admn)	Ahmad Salman
Public Relations Officer	Capt. N.M. Rashed
Director of Programmes	S.A. Hafeez
Station Director, Dhaka	Capt. Ahsanul Haque
Station Director, Karachi	Asnain Qutb
Station Director, Peshawar	Afzal Iqbal
Station Director, Lahore	Agha Bashir Ahmad
Director of News	Mohamad Sarfraz

Director of Staff Training	S.S. Niazi
Director of External Services and Monitoring	G.K. Farid
Research Engineer	S.A. Aziz
Maintenance Engineer	Bashir Ahmad
Engineer-in-charge HPTs	M.A. Rahman

Conditions soon began improving with the enthusiasm of the staff of the Radio Stations spurring the course of development. As against the administrative set-up comprising the Headquarters, three Regional Stations and three News Units in 1947, the position just before the 1971 dismemberment of the country was that there were 15 Broadcasting Houses –9 in West Pakistan at Karachi, Hyderabad, Lahore, Rawalpindi-I, II and III, Peshawar, Quetta and Multan and six in East Pakistan at Dhaka, Chittagong, Rajshahi, Sylhet, Rangpur and Khulna. There were 31 transmitters (17 medium wave and 14 short wave) with radiating power of 1120.5 kilowatts (543 KW MW and 577.5 KWSW).

After the fall of Dhaka, Pakistan was left with nine stations and 21 transmitters (10 medium wave and 11 short wave) with radiating power of 856 KW (396 KW MW and 460 KWSW). At the end of June 2000 there were 24 Radio Stations and 46 transmitters (27 medium wave, 14 short wave and 5 FM) with a total radiating power of 3854 KW (2701 KWMW, 1141 KWSW and 12 KWFM). The actual number of FM transmitters is 26 with radiating power of 27.63 KW. The remaining transmitters are used as STL. The power of medium wave transmitters ranges from 0.25 KW to 1000 KW, of short wave transmitters from 1 KW to 250 KW and of FM transmitters from 50 watts to 5 KW.

On the eve of the incorporation of Radio Pakistan there were 32 Units in addition to the Directorate General, which was divided into four parts. Of these 32 Units, 9 were Broadcasting Houses, 7 News Units and 7 HPTs. Other Units were the Central News Organization, External Services, Transcription Service, Commercial Service, Monitoring Department and Radio Publications all located at Karachi; Staff Training School, Technical Training School and Receiving Centre located in Islamabad. The total number of employees of Radio Pakistan in position at that time was 2417 (sanctioned strength =2512) and 377 staff artists.

Before incorporation each Station had a Regional Director in charge. In the discharge of his functions he was assisted by a Regional Engineer, two to six Assistant Regional Directors, a Listener Research Officer, a School Broadcast Officer, a Senior Radio Engineer, an Assistant Business Manager, Radio Engineers, Programme Organizers, a Presentation Organizer, Producers and several other technical and administrative staff.

Similarly each High Power Transmitting Station was run by a Resident Engineer, assisted by Deputy Resident Engineers, Radio Engineers and other technical, administrative and maintenance staff. A News Unit, except the Reporting Unit at Rawalpindi, usually consisted of six or seven persons including a News Editor, Assistant News Editor, and Sub-Editor.

Functions of PBC

Pakistan Broadcasting Corporation came into being on 20 December 1972. Its functions as laid down in the PBC Act of 1973 are:

a) To provide broadcasting services for general reception in all parts of Pakistan and the territorial waters thereof and on board ships and aircrafts and in other countries and places for the purpose of disseminating information, education and entertainment through programmes which maintain a proper balance in their subject-matter and a high general standard of quality and morality.
b) To broadcast such programmes as may promote Islamic ideology, national unity and principles of democracy, freedom, equality, tolerance and social justice as enunciated by Islam, discourage parochial, racial, tribal, sectarian, linguistic and provincial prejudices and reflect the urges and aspirations of the people of Pakistan.
c) To broadcast in the Home Services such special programmes as the Federal Government may, from time to time, direct.
d) To broadcast programmes in External Services to such countries and in such languages and at such times as the Federal Government may, from time to time, direct.
e) To bring to public awareness the whole range of significant activity and to present news or events in as factual, accurate and impartial a manner as possible.

f) To carry out instructions of the Federal Government with regard to general pattern or policies in respect of programmes, announcements and news to be put out on the air from time to time.
g) To hold the existing, and to construct or acquire and establish or install additional stations and apparatus.
h) To hold the existing, and to construct or acquire additional equipment and apparatus for telephony in Pakistan for purposes of broadcasting.
i) To compile, prepare, print, publish, issue, circulate and distribute, with or without charge, such papers, magazines, periodicals, books, circulars and other such matter as may be conducive to any of the functions of the Corporation and
j) To collect news and information in any part of the world in any manner that may be deemed fit.

The Corporation is guided by the Federal Government on questions of policy.

Board of Directors and Functions

A Board of Directors appointed by the Federal Government administers the affairs of the Corporation. They consist of a Chairman, a Director General who is the chief executive of the Corporation, and not more than six other Directors. The Board has full powers to prepare and approve the annual revenue budget of the Corporation. It formulates and implements all programmes and policies, and drafts plans for technical development within the country and for promotion of the Corporation's interests abroad.

The Corporation has the power to borrow in Pakistan currency or foreign exchange, and to draft regulations where necessary, with the previous sanction of the Federal Government.

At present PBC has five full-time Directors, each heading a Wing of the Corporation. Each Director is assisted by Controllers, Deputy Controllers, Programme/Engineering/Administrative/Accounts Managers or News Editors and other professional and ministerial staff. On its conversion into a Corporation certain structural changes were made. These included: (i) merger of the two training schools into one, (ii) making Monitoring Department a part of the Central News

Organization, (iii) redesignation of Commercial Service as Sales with head office at Karachi and Zonal offices at Rawalpindi and Lahore, (iv) merger of independent News Units at Broadcasting Houses with Stations, (v) redesignation of Transcription Service as Central Productions with head office at Rawalpindi and two units at Karachi and Lahore, (vi) opening of new departments/units like the Procurement, Construction, Legal and Management Facilities and (vii) creation of a new Finance and Accounts Wing with all its ramifications.

As a result there are now 55 Units in PBC including the Headquarters. These are 24 Broadcasting Houses, 3 Central Production Units, 3 Sales Units, and 13 HPTs. Other Units are: Central News Organization, External Services, World Service, Procurement Cell, Projects Cell (Two Units), Equipment Production Unit, Receiving Centre, Laboratory and Workshop, Pakistan Broadcasting Academy and Radio Publications.

The total number of posts in PBC at the end of June 2000 was 5350 excluding 463 posts of staff artists. Of these 1525 (28.5 per cent) were class IV and 1395 (26.3 per cent) professionals in PBC scale 5 and above. Due to a ban on recruitment and shortage of funds as many as 1856 (34.7 per cent) posts were lying vacant. This shortfall is not proportionate among different cadres and departments. The worst hit are the professional cadres as will be clear from the following table of posts in PBC scale 5 at the end of June 2000:

Post	Sanctioned Strength	Actual	Vacant No.	%
Producer	152	-	152	100.0
Language Producer	36	9	27	75.0
Sub-Editor/Monitor/Sub-Editor (Language) News Listener	100	14	86	86.0
Asst. Audience Res. Officer	4	2	2	50.0
Broadcast Engineer	191	135	56	29.3
Asst. Admn. Officer	36	23	13*	36.1*
Asst. Accts./Audit/Revenue Officer	56	48	8	14.3
Sales Officer	11	11	-	0.0
Store Officer	9	8	1	11.1
Others	30	17	13	43.3
Total	625	267	358	57.3

* = 15 posts of Asst. Admn. Officers were created during 1999-2000.

It will be seen that the incidence of vacancies in all but two of the professional cadres ranged between 75 per cent and 100 per cent while in the supporting cadres it was below 14.3 per cent. In the case of Assistant Administrative Officers, the 36.1 per cent vacancies shown are due to 15 new posts created during the year 1999-2000. As such it could be said that there was no vacancy in that cadre.

The functions of the directors are:

1. **DIRECTOR OF PROGRAMMES:** Overall responsibility for planning, production, content and presentation of all programmes broadcast in all Home and External Services as well as those produced by the Central Production Units, Press and Public Relations, Publications, Audience Research and all other programme matters in accordance with the instructions of the Director General and the policies and regulations framed by the Board of Directors.
2. **DIRECTOR OF NEWS AND CURRENT AFFAIRS:** Overall responsibility for collection, editing and presentation of news broadcast in all Services, planning, production, contents and presentation of current affairs and news bulletins, newsreels and monitoring department in accordance with the instructions of the Director General and policies and regulations framed by the Board of Directors.
3. **DIRECTOR OF ENGINEERING:** Overall responsibility for project implementation, installation, operation and maintenance of technical plants and equipment, technical planning, research and development, construction and maintenance of buildings, procurement of technical equipment and spares, security of engineering installations and all other engineering matters in accordance with the instructions of the Director General and policies and regulations framed by the Board of Directors.
4. **DIRECTOR OF ADMINISTRATION:** Overall responsibility for appointments, promotions, transfers, training and all other personnel matters of the officers and staff and staff artists of all departments of the Corporation, management facilities, training, overseas liaison, secretariat, security, legal matters, management information, and other administrative matters in accordance with the instructions of the Director General and the policies and regulations framed by the Board of Directors.
5. **FINANCE DIRECTOR:** Overall responsibility for procurement, supervision and operation of finances of the Corporation, control

of recurring, non-recurring and development expenditure, maintenance of accounts, preparation of budget estimates, internal and external audit, collection of radio licence fees, sale of advertisement time, income from other sources and all other financial matters in accordance with the instructions of the Director General and policies and regulations framed by the Board of Directors.
6. A post of Consultant with the status of Director was created to head the newly established Wing of External Services and Current Affairs in August 1999.

PBC Wings, Departments and Branches, their Personnel and Functions

I. Programme Wing: The Programme Wing has the following major sections:

1) Home Services,
2) External Services,
3) World Service,
4) Central Productions,
5) Press and Public Relations,
6) Publications,
7) Audience Research.

1) Home Services: The Section is headed by a Controller and assisted by Deputy Controllers and Programme Managers. The functions of the Section include:

a) Policy directives and guidelines to stations and their implementation.
b) Coordination and control of all types of programme, educative, informative and entertaining and their quality control.
c) Scrutiny of programme schedules of stations.
d) Fixed point charts and transmission hour schedules.
e) Anniversaries and Festivals and National Days of friendly countries.
f) Supervision of national hook-ups.
g) Coverage of VVIPs on their foreign visits.

h) Central and Regional Advisory Committees.
i) Station Directors' Conference.
j) Liaison with various wings/departments/sections within the Corporation and outside agencies.
k) Liaison with literary, cultural, sports and other specialized organizations/institutions.
l) All programme sanctions beyond the financial powers of Station Directors/Heads of Unit.
m) Fee structure and booking policy.
n) International sports coverage.
o) Listeners' letters and press comments and follow-up.
p) Exchange of programmes under cultural agreements.
q) Facilities to foreign journalists.
r) Central and Regional Qirat Committees.
s) Returns/Statements to Ministry of Information and Media Development and other allied agencies.
t) Overall administration of Home Service Branch.

The Home Service network comprises 24 broadcasting stations. Each Regional Station caters primarily for the requirements of the listeners in its own zone. Stations programmes are intended to (a) reflect the artistic, social and cultural traditions of the nation, (b) to impart up to date information on literary, artistic, cultural, social, economic, political and other subjects, (c) to bring out the common ideals and aspirations of the people of Pakistan, (d) to build up homogeneous outlook and singleness of purpose and to direct the energy of the nation into constructive channels and (e) to provide healthy entertainment. These purposes are served by arranging various categories of programmes in different languages for different sections of society.

Each Station is headed by a Station Director of the status of Controller, Deputy Controller or Programme Manager. Islamabad, Karachi, Lahore, Peshawar, Rawalpindi and Quetta Stations are headed by a Controller; Turbat, Sibi, Loralai, Zhob, Abbottabad and Chitral Stations by a Programme Manager while all the remaining twelve Stations are run by a Deputy Controller. The Station Director is assisted by one or more Programme Managers, Engineering Manager, News Editor, Administrative Officer, Accounts Officer, Sales Officer and Audience Research Officer who look after their respective branches. The Station Director is the administrative head of the Broadcasting

House and exercises financial and administrative powers as such. The main functions are:

a) To conceive, plan and implement new programme ideas.
b) To prepare quarterly schedules and submission of the same to the Headquarters. Schedules are prepared in accordance with the pre-determined ratios between national and regional languages, between classical, light and folk music, between spoken-word items and music, ratio of special audience programmes and general instructions of the Director General.
c) To cover functions/events of public interest in the region.
d) To feed daily programmes of current affairs.
e) To prepare and broadcast local news bulletins.
f) To send despatches to Central News Organization, Islamabad about items of national interest.
g) To scout new talent.
h) To maintain and operate all equipment at Broadcasting House and to coordinate between BH, HPT and Receiving Centre.
i) To maintain and physically verify technical stores and spare parts, etc.
j) To maintain and repair office and residential buildings, land, water supply and electrical installations.
k) To liaise and coordinate with the postal, telephone and telegraph authorities.
l) To deal with all administrative matters i.e. service record of the staff, recruitment, promotion, medical facilities, pension and gratuity, telephone and electricity, office equipment, transport, stationery, liveries, furniture, etc.
m) To deal with union and legal matters.
n) To keep liaison with advertisers and advertising agencies in connection with advertisements and sponsored programmes.
o) The Accounts Section in a Broadcasting House is looked after by an Accountant/Accounts Officer who is responsible for his work to the Controller (Finance and Accounts) but administratively to the Station Director.

2) External Services: Starting with four services on 14 August 1949 each with a duration of 45 minutes, PBC at the end of June 2001 broadcast seventeen external services in as many languages for a total duration of 12 hours a day. The External Services are headed by a Controller. The objectives of External Services include:

a) Projection of Pakistan abroad with special reference to its foreign policy, struggle for Pakistan, its economic development, social, cultural and literary activities.
b) Strengthening ties with Muslim and Third World countries in particular and with all other countries in general.
c) Supporting struggle for self-determination of subjugated people.
d) Countering any biased or hostile propaganda against Pakistan through dissemination of objective facts.

The Controller in the discharge of his functions is assisted by two Deputy Controllers, two Programme Managers, three Unit Managers, Senior Language Producers, Language Producers, etc.

3) World Service: The World Service was started on 21 April 1973 to keep the Pakistanis abroad particularly in the Gulf, Middle East, South Asia and Europe abreast of the activities of their country to create in them a sense of belonging and motivating them to develop a sense of involvement. It radiates programmes in five transmissions for a duration of ten hours a day. Besides informing overseas Pakistanis about developments in different fields the Service also broadcasts programmes to attract foreign investment by Pakistanis or foreigners.

The Service is headed by a Deputy Controller and assisted by a Programme Manager, four Senior Producers, three Producers, etc.

4) Central Productions: The Central Productions was set up in January 1960 under the style 'Transcription Service'. It has three Units located at Islamabad (head office), Lahore and Karachi. The Islamabad Unit is headed by a Controller while the other two Units are looked after by a Deputy Controller each, with the help of Programme Managers, Engineering Managers, Senior Broadcast Engineers, Broadcast Engineers, Senior Producers, Producers, etc. The objectives of the Section include:

a) Production of quality programmes for preservation and distribution to home and foreign stations.
b) Preservation of the cultural heritage of Pakistan through the medium of sound.

c) Production of special programmes for special occasions for broadcast from foreign and home broadcasting stations, and
d) Experimentation in new production techniques.

5) Press and Public Relations: The Head (Deputy Controller) of Press and Public Relations performs mainly the following functions:

a) To keep liaison with the press for giving news items regarding PBC activities and to reply to criticism in the press.
b) To publish annual reports, house journals and other books, pamphlets, cards, etc.
c) To do public relationing for maintaining goodwill of the Corporation.
d) To prepare annual report on PBC activities for submission to the Federal Government.
e) To arrange replies to questions in the National Assembly and the Senate about PBC.

6) Publications: The Radio Journals office was set up at Karachi in July 1948. The Publications Section is headed by a Deputy Controller and assisted by a Programme Manager, two Sub-Editors, Sales Officer, Sales Assistant, Proof Reader, etc. It is responsible for publishing two monthly programme journals: 'Ahang' in Urdu and 'Pakistan Calling' in Urdu, English, Farsi, Arabic, Dari and Turki. 'Ahang' has a large magazine section. 'Pakistan Calling' prints talks broadcast in the External Services on national topics and issues. The Section has published some useful books based on material broadcast from Radio Pakistan.

7) Audience Research: The Section is responsible for audience research to evaluate the effectiveness of the programmes and to obtain listeners' suggestions for possible improvements in programmes. The Section is headed by a Controller and assisted by two Audience Research Managers, eleven Audience Research Officers, four Assistant Audience Research Officers and nine Research Assistants. The Controller's services are being utilized in some other section. Most of the staff is posted at stations. There are only six persons at the Headquarters viz. two ARMs, one ARO, one AARO and two Research Assistants.

II. NEWS AND CURRENT AFFAIRS WING: On the eve of incorporation this Wing comprised Central News Organization at Karachi, Monitoring Department at Karachi, Reporting Desk at Rawalpindi and News Units at Hyderabad, Multan, Lahore, Peshawar, Quetta and Rawalpindi. The News Units were headed by a News Editor except the one at Rawalpindi which was headed by a Deputy Director of News who was assisted by four Assistant News Editors, one Sub-Editor and News Translators, etc. The entire set-up was manned by twelve News Editors, twenty four Assistant News Editors, five Sub-Editors, Language Supervisors and News Translators, News Listeners, etc. The Monitoring Department was being managed by two Assistant Directors with the help of nine Analysts, thirty Language Monitors, Regional Engineer, Technical Assistants, etc.

Immediately after incorporation the upper hierarchy of the Wing consisted of the Director of News, two Controllers, four Deputy Controllers, five Assistant Controllers, News Editors, Assistant News Editors, Sub-Editors, News Listeners, Monitors, Language Supervisors, News Translators, etc.

At the end of June 2000, besides the Director of News there were seven Controllers, twenty six Deputy Controllers, thirty one News/Monitoring Editors (of the status of Managers in other Wings), 10 Language Editors, 50 Assistant News Editors/Monitoring Officers/CNL, 11 Assistant Editors (Language), 65 Sub-Editors/Monitors, 33 Sub-Editors (Language), 8 News Listeners, etc. Of the seven Controllers, four are for News Section each being in charge of a shift. There is one Controller each for reporting, current affairs and monitoring. There is a News Branch headed by a News Editor/ANE at all stations. It works under the administrative control of the Station Director. There are thus four main sections of CNO:

1) News
2) Reporting
3) Current Affairs, and
4) Monitoring and Administration.

In August 1999 the Current Affairs Section was made part of a newly created External Services and Current Affairs Wing.

The main objective of the News Section is to collect and present news and events in as factual, accurate and impartial manner as possible. A total of 108 bulletins are broadcast daily. At present twenty foreign broadcasting stations are being monitored in six languages. On the basis of this material a Daily Monitoring Report is prepared. The Reporting Section coordinates the activities of the regional News Units and ensures coverage of important activities of the President, the Prime Minister, Senate/National Assembly, etc. The Current Affairs section is responsible for the broadcast of commentaries and discussions on events of political, economic and social importance.

III. ENGINEERING WING: On the eve of incorporation the Engineering Wing besides the Chief Engineer had six Deputy Chief Engineers, 16 Senior Engineers, 32 Regional Engineers, etc. At the end of June 2000 the Engineering Wing had besides the Director of Engineering, 12 Controllers, 44 Deputy Controllers, 112 Engineering Managers, 299 Senior Broadcast Engineers, 191 Broadcast Engineers, etc.

The Engineering Wing has the following Sections; each headed by a Controller and aided by Deputy Controllers, Engineering Managers, Senior Broadcast Engineers, etc:

1) Planning and Research Section
2) Frequency Management Section
3) Projects Section
4) Maintenance and Operations Section
5) Procurement and Supply Section
6) Equipment Production Section
7) High Power Transmitters and Receiving Centres

1) Planning and Research Section: The Section started functioning in August 1948. It is headed by a Controller and assisted by three Deputy Controllers, two Engineering Managers, two Senior Broadcast Engineers, etc. Its functions include:

a) Planning for development of broadcasting in the country with the object of optimum coverage with minimum cost.
b) Feasibility studies in respect of new stations, preliminary estimation of costs, preparation of PC-I proformas.
c) Processing of PC-I proformas with the Planning Division, CDWP and ECNEC.
d) Progress of development expenditure, processing of PC-III proformas with the Planning Division.

e) Selection of sites of new transmitters, planning of extension to old units.
f) Preparation and processing of PC-IV proformas showing particulars of projects completed and commissioned.
g) Advice on ADP requirements.

2) Frequency Management Section: The Section is in the charge of a Controller who is assisted by two Deputy Controllers, two Engineering Managers and seven Senior Broadcast Engineers. This Section was separated from the Planning and Research Section in 1988 to form an independent Section with the following functions:

a) Applied research, field intensity measurement, ground conductivity surveys, technical monitoring of spectrum on behalf of foreign organizations, spectrum analysis of medium wave bands, protection ratio in respect of frequency assignments both in medium wave and short wave.
b) Design and specifications of the new antennae for the required coverage.
c) Liaison with ITU, CCIR, IFRB in connection with adoption of international standards, plans and conformity with Radio Regulations.
d) International coordination with ITU in the usage and management of frequencies for minimum interference from other stations of the world.
e) Coordination with PSI, EBU and ABU in respect of international standards.
f) International hook-ups, liaison with Pakistan Telecommunication Corporation and other organizations for the same. Supervision of relays and national hook-ups.

3) Projects Section: In the beginning it was part of the Maintenance Section but was separated from that Department on 1 October 1953 and was called Installation Department. This Section has two Units at Rawalpindi and Karachi. The Rawalpindi Unit is headed by a Controller and assisted by two Deputy Controllers, 7 Engineering Managers, one Construction Engineer, 11 Senior Broadcast Engineers while the Unit at Karachi has two Deputy Controllers, two Engineering Managers and seven Senior Broadcast Engineers. It has the following functions to perform:

a) Detailed designing, estimating and technical specifications in respect of equipment.
b) Actual execution of work in the field for all projects according to ADP allocations.
c) Budgeting and control on capital expenditure including foreign exchange.
d) Testing and commissioning of all new equipment as per international specifications and handing over to the operating units.

4) Maintenance and Operations Section: This is the biggest Section of the Engineering Wing and was set up in August 1948. It is headed by a Controller who is assisted by four Deputy Controllers, four Engineering Managers and four Senior Broadcast Engineers for the following functions:

a) Technical maintenance and operation of all Stations/Units including HPTs, LPTs and Receiving Centres.
b) Detailed analysis of special breakdown reports to take corrective remedial measures. Watch over the performance of operating units.
c) Technical inspection of stations/units. Issuance of guidelines to technical heads for ensuring efficient performance of equipment and technical staff. Performance tests of equipment, machinery, etc.
d) Periodic review of maintenance schedule of stations/units, improvement, modification in the light of latest technical development for enhancing efficiency and prolonging life of equipment.
e) Special assignments to supervise important works at ll units as and when required or in cases of emergency or serious breakdowns technical help to stations/units; organization of special maintenance/repairs/modifications, if required.
f) Watch over the status and performance of power supply, telephone lines, etc. to all Units; Coordination, liaison with WAPDA and PTC; mutual coordination among different technical units of the Corporation.
g) Watch over the performance, stock position of all spare parts required for various units particularly performance of transmitting valves of all HPTs.
h) Repair and maintenance of all office and residential buildings and civil works at various units and the Headquarters.

i) Review and scrutiny of budgets of units on the engineering side; enforcement of financial control over technical expenditure.

5) Procurement and Supply Section: The Procurement Cell is headed by a Controller and assisted by three Deputy Controllers (including one from Accounts Wing), four Engineering Managers, one Store Manager, ten Senior Broadcast Engineers, etc. Its functions are:

a) Detailed scrutiny of lists for elimination of duplicity of requirements, economy of demand to a reasonable level.
b) Consolidation and preparation of indents of spare parts and capital stores.
c) Estimation of current prices. Comparison with previous prices. Determination of requirements of foreign exchange for the import of spare parts and of capital equipment. Preparation of spares purchase budget and equipment budget on capital side showing requirements of foreign exchange.
d) Registration and invitation of tenders. Determination of comparative merits of tenders. Finalization of orders and contracts.
e) Insurance cover, import licences, opening of L/Cs, processing of cases from the Import Relaxation Committee, clearance from the Customs.
f) Receipt and inspection of all consignments, warehouse examination reports, etc. Insurance surveys and claims.
g) Storage and preservation of equipment and stores till their final disposal. Maintenance of all stores' books and records.
h) Physical control and stocktaking.

6) Equipment Production Unit: The Equipment Production Unit was created in 1975 for producing locally most of the electronic equipment required by the Corporation. It is headed by a Controller with three Deputy Controllers, five Engineering Managers, 10 Senior Broadcast Engineers, etc. Its functions include:

a) Scrutiny of the specifications of equipment required by the Corporation with a view to producing it with the optimum use of locally available components.
b) To manufacture/fabricate as many parts as possible of electronic equipment in its own laboratory.

The entire recurring budget of this Section is said to be met from the cost of the equipment produced by it.

7) High Power Transmitters and Receiving Centres: There are thirteen HPTs at present. Each HPT is headed by a Deputy Controller except two at Faqeerabad and Rewat Complex which have a Controller as head. Each head is responsible for:

a) Operation, maintenance and efficient handling of breakdowns on transmitters and other equipment.
b) Processing requirements of spare parts.
c) Maintenance of office and residential buildings.
d) Electrical and water supply installations.
e) Security and safety of installations.
f) Liaison and coordination with TCP, PPWD, Police and WAPDA authorities.

A Receiving Centre normally receives programmes of national hook-ups, news and other relays and feeds them to the Control Room of a Broadcasting House. The BH feeds these programmes back to Transmitters. The Receiving Centre at Islamabad is used mainly for monitoring. The Monitoring Section of the Central News Organization is dependent on this Receiving Centre for good quality reception of foreign broadcasts.

IV. ADMINISTRATION WING: On the eve of incorporation the Administration Wing was headed by a Deputy Director General, who was from the Programme Wing, and assisted by one Director of Administration and four Administrative Officers—two at the Headquarters at Rawalpindi, the third at Headquarters at Karachi and the fourth in Central News Organization, Karachi. The incorporation of Radio Pakistan increased the volume of administrative work. The services which were earlier being provided by various Government Departments became the responsibility of PBC. Maintenance of accounts, internal audit; procurement of stationery, stores and equipment; provision of accommodation and medical facilities are some of the functions which have been taken over by the Corporation. Likewise all recruitment which used to be looked after by the Federal Public Service Commission is now handled by the Corporation. Similarly construction and maintenance of buildings and legal affairs were no longer the responsibility of Pak. PWD and the Law Division respectively but of PBC. This addition to the duties and functions of the Corporation as well as expansion plans created a number of job opportunities in the Corporation.

Immediately after incorporation the post of Deputy Director General (Administration) was transformed into that of Director of Administration and Public Affairs. He had eight sections under him, each headed by an Assistant Controller except the Staff Training School whose Principal was of the status of Deputy Controller and was assisted by two Assistant Controllers. The Sections were:

1) Personnel
2) Management Facilities
3) Liaison and Training
4) Legal and Secretariat
5) Staff Training School
6) Publications
7) Listener Research, and
8) Press and Public Relations

The last three Sections were later passed over to the Programme Wing. At the end of June 2000, the Administration Wing had the following four Sections:

1) Personnel Section
2) Administration Section
3) Training and Overseas Liaison Section
4) Pakistan Broadcasting Academy

The Personnel and Administration Sections are manned by two Controllers, five Deputy Controllers (including one Chief Security Officer), nine Administrative Managers, two Administrative Officers and fourteen Assistant Admn. Officers. The work of the two Sections is divided into branches as under:

i) Recruitment Branch
ii) Legal Branch
iii) Personnel Branch
iv) Coordination Branch
v) Pension Branch

vi) Establishment Branch
vii) Management Facilities Branch-I & II
viii) Regulation Cell
ix) Stores Branch

1) Personnel Section: The Personnel Section is headed by a Controller and assisted by three Deputy Controllers, four Administrative Managers, two Administrative Officers and three Assistant Admn. Officers. Functions of the Personnel Section include:

a) Recruitment of all categories of staff.
b) Promotions, transfers, deputations, secondments, seniority and confirmation.
c) Training for promotion and for apprentices.
d) Pay and allowances, increments, pay fixation, honorarium and leave cases.
e) Organizational structure, reorganization and creation of posts.
f) Disciplinary cases of all employees, their suspension, termination and resignation, appeals and representations, legal cases, contracts and union matters. Declaration of assets.
g) Regulations, organization and methods. Keeping Service Manual up to date.
h) Maintenance of ACRs (Annual Confidential Reports) and quantifiation of PERs (Performance Evaluation Reports).
i) Management information and statistics.
j) Assembly/Senate questions.
k) Coordination.

2) Administration Section: The Administration Section is apexed by a Controller and assisted by one Deputy Controller (Management Facilities) and one Chief Security Officer (Deputy Controller), two Administrative Managers, four Assistant Admn. Officers and one Stores Officer. The functions of the Section include:

a) To provide management facilities regarding: renting of accommodation for offices and residences, office furniture, stationery and other equipment, liveries, transport, telephone and medical facilities.
b) To deal with such matters as group insurance, benevolent fund, GP/CP funds and advances.

c) Staff welfare, pension and gratuity, service record of Headquarters staff, pay fixation, tour sanctions and TA/DA.
d) Maintenance and cleanliness of National Broadcasting House and garden, security and firefighting arrangements.
e) Arrangement of functions/parties.
f) Receipt and issue of *dak*, (official correspondence through post offices or messengers).
g) Maintenance, purchase and physical verification of stores and inventory.

The Chief Security Officer acts as adviser to the Director of Administration under the administrative control of the Controller of Administration, on all matters relating to security arrangements at Headquarters, Units/Stations in general and key points in particular. His functions include:

a) To check the suitability of existing security arrangements and advise heads of units/stations for remedial action.
b) Issue latest orders regarding security and firefighting.
c) Supervise all security staff.
d) Keep close liaison with relevant government authorities and departments.

For the discharge of his functions the CSO has a large staff spread all over the country. These include Security Officers, Security Assistants, Studio Commissionaires, Security Guards and Studio Guards.

3) Training and Overseas Liaison Section: In Radio Pakistan this function was handled by the Director of Liaison who was responsible for Press, Public Relations and Publications also. In the Corporation the Deputy Controller (known as Head TOL) who heads this Section is directly responsible to the Director of Administration. He has the assistance of an Assistant Admin. Officer, two Assistants and other ministerial staff. His functions include:

a) To look after the training programme in Pakistan Broadcasting Academy, some specialized agencies in Pakistan and in foreign countries.
b) To look after all correspondence and documentation for visits abroad of PBC personnel for coverage of President's and Prime

Minister's tours and international sports events, and for PBC officials going abroad to seminars, conferences, etc. and to pay goodwill visits under cultural exchange programmes.

4) Pakistan Broadcasting Academy: The affairs of Pakistan Broadcasting Academy are run by a Principal (Controller) with the help of three Vice Principals (Deputy Controllers)—one each from the Programme, Engineering and News Wings. In addition, there are five Engineering Managers, three Programme Managers, one News Editor, three Senior Broadcast Engineers and one Senior Producer. The supporting staff includes Broadcast Engineers, Electrical and Mechanical Instructor, Air-conditioning Supervisor, General and AC Mechanics, etc. The Academy is meant:

a) To organize basic induction courses and promotion courses for engineering, programme and news staff.
b) To organize special courses and symposia and seminars as and when required on different formats and services.
c) To organize special courses in collaboration with other Government and autonomous agencies to integrate broadcasting with development programmes, as well as in cooperation with foreign broadcasting organizations.
d) To ensure good and efficient performance of all training equipment, transmitters, studios, receiving centres and other technical areas at the Academy.
e) To evaluate the performance of trainees, finalize results for submission to the Headquarters.

V. FINANCE AND SALES WING: This is a post-Corporation creation. In Radio Pakistan, one of the Admn. Officers at the Headquarters among other things was also responsible for initiating, preparing and finalizing budgets and accounts, preparing all sorts of bills and vouchers, TA/DA, GP Fund advances, etc. with the help of an Accountant (UDC with special pay), one cashier (LDC) and one or two clerks. At Stations/Units this work was handled by an Accountant (UDC), one or two clerks including the cashier.

This Wing has two main Sections: Finance and Account Section and the Sales Section.

1) Finance and Accounts Section: The Finance and Accounts Section is manned by two Controllers assisted by four Deputy Controllers, nine Accounts Managers, 26 Accounts/Audit Officers, 55 Assistant Accounts/Audit Officers, etc. This Section has three main Branches:

a) Accounts and Finance Branch
b) Projects/Capital Branch
c) Audit Branch

The Accounts and Finance Branch is responsible for the preparation of all vouchers and bills, maintenance of Account Books, maintenance of individual fund accounts, and preparation of annual revenue and capital budgets. The Capital/Project Branch is responsible for financial control over the funds allocated to new projects in the ADP/Five Year Plans. The Audit Branch sees and checks that a proper system of internal financial and accounting procedures is fully followed. The functions of this Section are:

a) To prepare revenue and capital budgets as well as the foreign exchange budget for submission to the Board of Directors and the Federal Government.
b) To submit quarterly reports to the management about the financial affairs and progress made towards targets fixed in the budget.
c) To control financial performance of all Units/Stations through monthly budgetary control statements and statement of affairs.
d) To devise ways and means, systems and procedures for internal control and automatic check in day to day working.
e) To carry out reappropriation of finances from certain heads to other heads.
f) To control internal audit of accounts of units and apprise the management of the state of affairs at units.
g) To release funds to units monthly to meet budgeted expenditure and other genuine requirements.
h) To prepare cash flow statements for submission to the Federal Government and PBC Management.
i) To prepare balance sheet and profit and loss accounts.
j) To arrange for the conduct of external audit by a firm of Chartered Accountants, and submit audited report to the Board of Directors.
k) To maintain accounts according to established commercial practices.

l) To arrange LCs, opening of bank accounts and other bank facilities required from time to time.
m) To analyse and scrutinize the financial implications of any project, or a proposal/demand of union/employees or of any new facility to be extended.
n) To maintain GP Fund and CP Fund accounts of employees and invest excess funds in safe securities, and
o) To prepare working papers and to correspond with Ministry of Finance for the grant of additional subsidy from the Government.

2) Sales Section: On the eve of incorporation the position of commercial service was that a Director looked after it with the help of a Programme Organizer, Programme Producer, Business Manager, Radio Engineer and others at its office at Karachi. There was one Assistant Business Manager posted at Hyderabad, Multan, Lahore, Rawalpindi and Peshawar.

After conversion of Radio Pakistan into a statutory Corporation a Controller was put in charge of this Section. Its Head Office is located at Karachi and two zonal sales offices are at Lahore and Islamabad, while a Sales Officer is posted at all major stations. The zonal offices are headed by a Deputy Controller. The basic objective of the Sales Section is to promote sales through the sale of air time for advertisements on Radio Pakistan. On the whole the three Sales Units are manned by one Controller, three Deputy Controllers, three Sales Managers, five Sales Executives, eleven Sales Officers, Audit Officers, Accounts Officers, Assistant Accountants, Assistant Audit Officers, Traffic Assistants, Sales Assistants, Commercial Spot Checkers, etc.

13

Finance

At the outset, Radio Pakistan was considered a government department and its expenditures were met from the budget of the Federal Government. The income from Broadcast Receiver Licences became part of the Government revenues as did the income from advertising. This position changed with the incorporation of Radio Pakistan.

Sources of Income

The principal sources of income of the Corporation are advertisements and Government grants. Broadcast receiver licence (BRL) fee was also a source of income that was, however, abolished from 1 July 1999. This fee was collected by the Pakistan Post Office Department as an agency to PBC under the Wireless Telegraphy Act, 1933. The amount collected by PPO was transferred to PBC after deduction of collection expenses and commission. This was PBC's income. The overall trend in BRL fee collection showed a steady increase in revenue in pre-Corporation days. After incorporation the trend continued upwards to 1979-80 when it touched the highest figure of Rs.38.825 million. Thereafter, there was a decline to the lowest figure of Rs.9.858 million in 1998-99.

The rise in the amounts in the post-Corporation first seven years could be attributed to effective collection measures taken by PBC and PPO and increase in the rates from Rs.10 per domestic licence in pre-Corporation period to Rs.15 on 1 January 1973 to Rs.20 from 1 July 1975 and later to Rs.30 from 1 July 1990. The rates of different types of Broadcast Receiver Licences are given below:

Type of BRL	Before Corporation	From 1.1.1973	From 1.7.1975	From 1.7.1990
Domestic	Rs.10	Rs.15	Rs.20	Rs.30
Car	-	-	Rs.30	Rs.60
Commercial	-	Rs.50	Rs.75	Rs.150
Dealer	-	-	Rs.100	Rs.160

The number of licences issued and renewed had been erratic but on the whole rose from 1.57 million in 1972-73 to 1.80 million in 1979-80. Thereafter, it dropped to 0.42 million in 1998-99. PBC took a number of measures to boost income from this source. A proposal was made to amend the Wireless Telegraphy Act of 1933 to take control of BRL from PPO and devise a better system of collection of licence fees to plug the leakage in income and hence improve the financial position of the Corporation. PBC had not been successful in carrying this out and took measures such as the radio prize scheme of 1987 to boost license revenues. According to this scheme, listeners who obtained or renewed Radio licences during the quarter March-May 1987 were to be awarded the following incentives:

First Prize: Six return tickets for *Umra* through ballot.
Second Prize: Free *Ahang* for one year to 100 licence holders through ballot.
Third Prize: Free licence for annual subscriber of *Ahang* during the quarter.

In addition to Post Offices, licenses were issued at Broadcasting Houses as well. However, these measures failed to bring in larger revenues from renewal/issuance of licences. From 1 January 1990, postmen were authorized to issue or renew radio licences at the doorsteps of radio listeners. The postman was given an incentive of Re.1 per license in addition to a PPO share. This pilot project of PPO also failed to increase subscriptions. To urge listeners to get a license it was announced in January 1990 that only those request letters to Radio Station's programmes would be entertained which contained the number and date of issue of BRL. Popular comperes and stock characters appealed to listeners on air to get licensed.

The various schemes launched from time to time to augment income from Radio licence fees failed for three main reasons. First, it was difficult to trace unlicensed radio sets; secondly, Radio Inspectors as employees of PPO, did not seek out defaulting persons; and thirdly PPO was not affected by a rise or fall in BRL income as much as PBC hence their disinterest in solving the problem. PBC on the other hand was concerned with the fall in licence fee revenue due to rising financial costs of its ever-expanding activities.

Surprisingly, while PBC income from BRL was dwindling year by year, PPO's share had been rising since incorporation. For example, in 1973-74 (first full year after incorporation), out of the gross collection of Rs.23.43 million, PBC got Rs.20.97 million (89.5 per cent) while PPO's share was Rs.2.46 million (10.5 per cent). As against this, during 1998-99, of the gross collection of Rs.28.656 million, PBC share was Rs.9.858 million (34.4 per cent) and that of PPO was Rs.18.798 million (65.6 per cent). In other words to get 34.4 per cent of license fee revenue PBC had to surrender 65.6 per cent of the total to PPO, which according to financial experts was a bad tax/levy situation that was best done away with. To look at it from another angle, during the 26 year period from 1973-74 to 1998-99, the share of PPO spiralled by 664 per cent while that of PBC actually declined by 53 per cent, which was indicative of something seriously wrong with the sharing formula.

ADVERTISEMENTS

At the present time, PBC's primary source of income is sale of air time. The income from advertisements has shown a steady increase from year to year from the beginning with some minor exceptions, e.g. during 1965-66 as a direct result of the September war with India. From 1973-74, the business contracted rose from Rs.10.20 million to Rs.90.36 million in 1999-2000. This significant improvement can be attributed to concerted efforts made by the Sales staff, periodic raise in rates and availability of more air time for sale due to increase in programme hours and opening of new stations.

In order to provide an incentive to Sales staff and to advertising agencies, PBC introduced the concept of cash and commercial awards. All members of the Sales Section who meet annual targets are given a cash award of Rs.2000/-. Commercial awards were introduced in the year 1987-88 to acknowledge advertisers and advertising agencies for their business. The awards were presented to the first and second positions in five categories. In addition an award was given for the Client of the Year and another for the Agency of the Year. An awards function was held on February 3, 1989 in a Karachi hotel. During the year 1987-88, PBC's income from advertisements soared to over Rs.52 million. Thereafter, except for the years 1990-91 and 1991-92, it showed a rising trend reaching Rs.90 million during 1999-2000.

GOVERNMENT GRANTS

The other source of PBC income and its main source of revenue is the grant-in-aid it receives from the Government. The Government grant has gone up from Rs.26.06 million in 1973-74 to Rs.761.99 million in 1999-2000. This grant is meant to bridge the gap between income and expenditure as provided in the PBC Act, 1973. The grants are finances for commercially non-viable projects undertaken at the behest of the Federal Government. These grants-in-aid accounted for 79 per cent of the total income during 1995-96 and 87 per cent in 1999-2000 against 45 per cent during 1973-74. This reflects PBC's growing dependence on Government grants. In addition to these grants-in-aid, the Federal Government gives development grants each year for projects including new stations, transmitters, modernization and replacement.

EXPENDITURE

As compared with an increase of 1406 per cent in the total income of PBC from 1973-74 to 1999-2000, the total operational expenditure of the broadcasting service went up by 1478 per cent in the same period from Rs.56.36 million to Rs.889.20 million. The expenditure of the Corporation has generally remained under control and increased by 4 to 7 per cent a year to absorb annual increments of staff, salaries and price hikes, except in years during which there was a revision in pay scales, increased power costs, expansion projects, etc. The highest increase in a single year was during 1974-75 when total expenditure went up 40 per cent over the previous year mainly due to a union agreement and new pay scales.

The single biggest item of expenditure is salaries and allowances that normally account for about half the annual expenditure figures. Pension and gratuity, personal expenses, travelling expenses and other administrative costs consume about one-fourth of the total budget, bringing the total to about 75 per cent on administration costs. Power and fuel charges are the second biggest item of expenditure after salaries and allowances and generally account for more than 10 per cent of the total expenditure.

Measures to Increase Resources

PBC's attempts to increase income resources and curtail expenditure have for the most part not met with success. Every year, the budget has been reviewed in an attempt to do so. One idea was to increase income from BRL by revising and rationalizing the sharing formula or by hiring a private collection agency. One proposal suggested adding a small radio fee to the electricity bill as has been done in certain countries.

Another idea was that Government departments and agencies and autonomous bodies should provide services to PBC on concessional rates on a reciprocal basis, as is currently being done by PBC on a unilateral basis. A reciprocal arrangement would be a fair one and would ease PBC's financial crisis. If this is not agreed to, PBC may have to do away with its concessional rates to these parties as well. This would mean either a cut of about 10 to 30 per cent in expenditure of power and fuel, travelling, telephone, medical, etc. Alternatively, there would be an almost equal increase in advertisement income.

A better measure of permanent nature would be to drastically curtail or freeze expenditure on supporting/auxiliary services, which has gone up phenomenally since incorporation. The increase has been out of all proportion to rational expenditure increases and against the announced objectives of the Corporation. It has resulted mainly from the creation of unnecessary and superfluous posts at all levels particularly in the higher grades of 5 and above, and most of these posts are redundant in nature. This point gets more attention in the last chapter, The Future.

14

COMMERCIAL SERVICE/SALES

Commercial broadcasts have been a part of broadcasting history in the subcontinent from the very beginning. Private broadcasting enterprises as far back as 1923 depended on advertisements for their revenues. The Indian Broadcasting Company went commercial in 1927. It financed itself through advertisements and sponsored programmes. This continued till 1930 when the Indian Government took over and set up the Indian State Broadcasting Service. The commercials were left out by ISBS (later AIR) because the Government thought that broadcasting was primarily a social service particularly in a developing country.

This concept of social service was transferred to Radio Pakistan as a sacred legacy of broadcasting. It was fourteen years later that Radio Pakistan began commercial broadcasts from Karachi on 1 November 1961. This was during the midday transmission when one hour commercials were played on an experimental basis with the object of publicising locally manufactured products, thus helping the campaign of 'Patronize Pakistan Products'. Mr.Shamsuddin Butt was the first Director, Commercial Service.

There was a great rush to book commercial spots and the entire allocated advertising time was booked leading to demands for more advertising time. The success of commercial service from Karachi led to the opening of a commercial unit at Lahore Radio Station in 1967. On the demand of advertisers it was decided to start commercial service from Hyderabad, Rawalpindi and Peshawar in 1968. Later it was extended to other stations. World Service followed suit in 1974 and the External Services a year later. Some of the dates of starting Commercial Service at different stations are:

Karachi	1 November 1961
Lahore and Dhaka	1 May 1967

Rawalpindi, Hyderabad, Peshawar and Chittagong	1 July 1968
Multan and Quetta	1 March 1972
Bahawalpur	1 July 1975
Islamabad	5 July 1976
World Service	1 July 1974
External Services	1 July 1975

During 1964-65, the duration of commercial broadcasts was three hours and ten minutes on week days and 3 hours 40 minutes on Sundays. Over 300 products and services were advertised. In 1968 the duration was 95 hours a week i.e. 13 hours 35 minutes a day. The commercial revenue progressively increased. During the national emergency of 1965, regular programmes of this Service were suspended and were replaced with patriotic songs and war-related features and news. This caused a fall in advertising income. From 1 March 1966 normal commercial programmes were again started.

From 1 April 1973 commercials and sponsored programmes were permitted during the morning and evening transmissions. At Karachi the commercial service was so heavily booked that it was no longer possible to accept more commercials. On most days 2 minute 30 second commercial sessions were scheduled for every 3 minutes of recorded music. With commercial breaks extended to morning and evening transmissions the situation was eased. Similarly at Lahore Station commercial business was brisk. Today the full day's transmission at all Stations is open for commercials except the news and religious segments.

For commercial broadcasts the basic idea is to convey a message or introduce a service to listeners in an effective manner. This is a technical job and is handled by the Sales staff. Commercial programmes are of three types—sponsored programmes, spots (slogans/messages) and jingles. Sponsored programmes are usually of not less than five minutes and consist of narration, compering or music. These programmes are prepared either by the client or by PBC according to the policy and code of ethics prescribed for commercial broadcasts. Spots are normally in spoken words and of not less than seven seconds duration. A jingle is essentially a message or slogan set to music.

The head office of the Sales Section is located at Karachi and is managed by a Controller. It has two Zonal Sales Offices at Islamabad and Lahore, each headed by a Deputy Controller. At other Stations there is a Sales Manager or Sales Officer.

FUNCTIONS OF SALES SECTION

The Sales Section is responsible for promoting the sale of air time and sponsorship of programmes. It produces printed material, Radio Code/Guide, promotional brochures, leaflets, greeting cards, etc. Coordinates with advertisers and advertising agencies and performs a host of other duties related to the sales and marketing function.

CODE OF ETHICS

Before each item goes on air it is scrutinized to ensure that it does not contain any material which is against the ideology of Pakistan and the policies of the Government; is not libellous or defamatory, or morally inappropriate; factually incorrect; and likely to create sectarian, racial or regional hatred. The responsibility for the scrutiny and vetting of all broadcast material rests with PBC. The aim is to ensure that the normal code of PBC is not violated nor are the standards of decency and propriety compromised. Broadcasts likely to disturb public peace and of a libellous nature are not permitted. There is a ban on the sponsorship of purely religious and political broadcasts and on exploitation of religious or political subjects or personalities for commercial purposes. Biased opinions on trade or industrial disputes and advertisements appealing to racial/communal hatred are not allowed.

Special care is taken to avoid misstatement of facts as well as possible deception through implication or emission. No announcement is permitted which makes a claim that cannot be substantiated and without further qualification. Also no disparagement of competing goods is allowed. The consumer is regarded not merely as a purchaser of merchandise but as a thinking, discriminating, open-minded person who values decency and honesty. Advertisements regarding pharmaceutical products are required to be cleared by the Ministry of Health. Most of the commercial items are contracted at Lahore or Karachi where a Committee scrutinizes these items before sending to stations. Individual Station Directors again vet these items before putting them on the air.

National welfare and 'first obligation to the listener' have always been the primary concern of Radio broadcasters, while monetary considerations have been of secondary importance. This becomes very obvious when compared with the example of Television, which is a highly commercial-oriented medium. Many creditable programmes on TV are normally not produced or scheduled unless sponsored. An unsponsored programme is scheduled in unpopular time slots irrespective of its importance for viewers. All programming, even the Azan is adjusted according to the convenience of sponsored programmes. On the other hand with PBC, programmes for public benefit or of general interest are scheduled and broadcast at prime time often at the cost of sponsored programmes. This is an important fact that should be considered when assessing the revenue earning efforts of PBC.

There are conflicting views about the usefulness of commercial broadcasts. Some critics think that advertisements create an artificial demand for goods and services that are not necessary in life, leading to frivolous extravagance and wasteful spending. Others argue that advertising does nourish the consuming power of man but it also raises the standard of life. It sets goals of better lifestyles and spurs individual exertion for greater production. The American high standard of living is fuelled by the imaginative genius of advertising that not only creates demand but also by its impact upon the competitive process, stimulates the individual in quest of improving the quality of products.

ALLAMA IQBAL OPEN UNIVERSITY PROGRAMMES

Radio is the ideal medium in a developing country for the spread of formal education and mass literacy. In addition t its own educational programming Radio Pakistan also broadcasts programmes of Allama Iqbal Open University at concessional rates. The first AIOU programme was put on the air on 4 October 1976 and PBC charged only 25 per cent of the usual rates. These were broadcast from seven or eight stations. At present Jama Nama (University Newsletter) goes on air from Islamabad, Karachi, Hyderabad and Quetta while the lessons are carried by Islamabad Station only at the reduced rate of 33 per cent.

15

Central Productions

Transcription Service

The idea of having a Transcription Service was mooted as early as 1949. When the Staff Training Centre was set up that year it was to handle the proposed Transcription Service. Before this idea could materialize, however, the steady growth of the programme activities of Radio Pakistan came to a halt in 1953 as a result of prevailing national economic conditions. Radio's budget allocation for 'Allowances to Artists' was reduced by 40 per cent causing a severe set-back to the organization.

No broadcasting organization can afford to cut back to previous lower standards and disappoint its audiences. Thus the aim was to remain within the financial limits set and still ensure an acceptable and efficient programme service while retaining existing transmission hours. The answer lay in starting a Transcription Service. It was to be different from the Transcription Service of other major broadcasting organizations of the world that are better funded and consequently can afford better programming. Radio Pakistan was constrained by austerity measures.

The Transcription Services was based on off-the-air recordings of the music and dramatic programmes originated by Karachi and Lahore. Even Karachi and Lahore stations could not broadcast music and dramatic programmes throughout the month, but had to broadcast live programmes on alternate fortnights and share recordings for the other fortnight. Rawalpindi and Peshawar Stations were not to originate any general music and dramatic programmes, but received recordings made available to the Transcription Service by Lahore and Karachi.

This Transcription Service enabled Radio Pakistan to retain its existing transmission hours and its existing categories of programme and still remain within its financial limits. Although a great number of artists and performers were deprived of their main source of livelihood, the 'Transcription Service' helped Radio Pakistan to tide over the difficulty it was confronted with at the start of the year. The Service was responsible, in a period of ten months, for circulating recorded programmes of well over one thousand hours duration—an achievement unprecedented in the history of broadcasting.

This was, however, a temporary measure supervised by the Director of Music. More than a decade after the idea of Transcription Services was formulated in 1949 did it materialize in January 1960, though preservation of selected items of national importance on disc had started much earlier. The disc recordings of Quaid-i-Azam's speeches is one such valuable example. The Transcription Service started with a Director, two Programme Organizers, one Regional Engineer and four clerks in two rooms over a garage without a studio and at a distance of about four furlongs from the Karachi Station. In appreciation of regional requirements of East Pakistan, a Transcription Unit was set up in the Dhaka Station of Radio Pakistan in 1966. Its purpose was to capture in sound the life of Pakistan, touching on every aspect of activity for preservation and broadcasting.

FUNCTIONS

The Transcription Service produced quality programmes for distribution to stations and to broadcasting organizations of other countries on special occasions and on request. It was responsible for the preservation of the cultural heritage of Pakistan through the medium of sound. It also produced long-playing records, cassettes of music items and feature programmes in cooperation with EMI/Shalimar Recording and Broadcasting Company and other agencies and was responsible for the sale of recorded material.

When Radio Pakistan was converted into a Corporation the name of Transcription Service was changed with effect from 5 March 1973 to Central Productions and three Units were set up at Islamabad, Lahore and Karachi. The Islamabad Unit acts as head office and is headed by a Controller while Karachi and Lahore Units are managed by Deputy Controllers. In July 1973 the tape library was shifted to Rawalpindi where it was organized on scientific lines.

The Karachi Unit concentrates on spoken-word items while the Lahore Unit specializes in music recordings. It has a full-time orchestra attached to it. The head office at Islamabad looks after the National Sound Archives, coordinates and supplies recorded material to Stations within the country and broadcasting stations/organizations in foreign countries free of charge if they wish to put them on the air on their national network. All items have been card indexed with two or three cross-references. Complete catalogues of the recordings are available.

SOME ACHIEVEMENTS

The CPU has done some very useful and valuable work during the forty years of its existence. Briefly it has produced:

1) Special programmes on Pakistan Day, Independence Day, Defence of Pakistan Day, Iqbal Day, birth and death anniversaries of the Quaid-i-Azam Mohammad Ali Jinnah, Liaquat ali Khan etc. National and patriotic songs to suit the occasions were also specially got from poets and recorded.
2) Curtain-raisers on Pakistan and other countries on visits of Pakistani VVIPs in English, Urdu and some other languages of the countries of the visit and also on the eve of similar visits by foreign Heads of State and Government to Pakistan.
3) Recording of light music items to replace or at least reduce the quantum of film music.
4) Recording of theatre songs of 1,620 minutes duration.
5) Features/plays based on music recorded by CPU.
6) Preservation and propagation of classical music. Over 250 *raags* were recorded.
7) Recording of the Holy Quran in the voice of seven *Qaris* including two Egyptians and Maulana Ehtishamul Haq, Qari Ghulam Rasool and Qari Khushi Muhammad Al-Azhari.
8) For Islamic Summit Conference special programmes including a special signature tune, recording of Iqbal's *Shikwa Jawab-i-Shikwa* in the voice of Noor Jahan and its Arabic version in the voice of Umme Kulsum.
9) Special programmes on the occasion of centenaries of the Quaid-i-Azam and Allama Iqbal and 700th anniversary of Amir Khusrau. A set of three cassettes containing all the available speeches of the Quaid-i-Azam under the title 'Forever to Remember' was produced on his centenary. Programmes under the title 'Naqsh-i-Dawam' in connection with Quaid-i-Azam's centenary.

10) Research in and preservation of all traditional folk music of different regions of Pakistan which provides a vital link in the evolution of music. Otherwise this well-knit rich socio-cultural heritage would have been lost to posterity.
11) Obtaining, recording and preserving material of historical importance in different fields of activity.
12) Recording of interviews of leaders and workers of Pakistan Movement, VVIPs, scholars, writers, historians, educationists, poets, critics, scientists, *ulema*, sportsmen, etc.
13) Recording of community songs, songs for industrial labour and workers, thematic songs.
14) Preparation of special programmes to celebrate silver jubilee of Radio Pakistan in August 1972. A total of 60 musical and spoken-word programmes were prepared, recorded and sent to all Stations for broadcast. Some of the programmes were written by Z.A. Bokhari and recorded in his voice.
15) Preparation and broadcast of special programmes to celebrate silver jubilee of Transcription Service/Central Productions in December 1985.
16) A 30-minute feature entitled 'When Spring Comes to Pakistan' produced in seven languages and supplied to 36 countries for broadcast on Pakistan Day in 1970.
17) Preparation of a casket containing six cassettes entitled 'Sounds of Pakistan' along with its coloured brochure, for presentation to visiting Heads of State/Government, dignitaries and heads of foreign broadcasting organizations.
18) Compilation of an anthology of selected Urdu and Persian poetry. It is an important contribution of Radio Pakistan to the realm of literature.
19) Production of a series of plays comprising eight episodes entitled 'Emergency Ward' dealing with socio-economic problems.
20) Producing in 1993 a series of plays 'Sachi Kahanian' containing 13 episodes on narcotic control for broadcast from Radio Pakistan Stations.
21) Producing 47 Arabic lessons for broadcast from Home Stations and the World Service.
22) Preparing lessons for teaching the recitation of the Holy Quran entitled 'Aaiye Quran Sharif Parhen'. The lessons are regularly broadcast from all stations.

Productions. Material is supplied to foreign broadcasting networks including ethnic radio stations SAARC programmes are exchanged within the member countries and their broadcast is arranged on national hook-up during each calendar month. The main feature of the anniversary programmes is the recording of special seminars, talks, discussions and feature programmes for broadcast from all PBC stations. Moreover, on national days special feature programmes are produced in eight foreign languages viz. English, French, Arabic, Farsi, Afghan-Persian, Spanish, Turkish and Indonesian and sent for broadcast to as many as 69 countries.

This treasure house of our national, spiritual and cultural heritage known as the National Sound Archives had as at 30 June 2000 recorded material of over 1.013 million minutes duration on about 24,000 tapes under the following broad classifications:

a) Recitation of the Holy Quran,
b) Religious Programmes,
c) *Hamds*, *Naats* and devotional music,
d) Voices of national and international leaders,
e) Series of talks on different topics,
f) Detailed interviews of leaders and workers of Pakistan Movement,
g) Plays,
h) Features,
i) Documentaries,
j) Music: Folk, Regional, National, Classical, Theatre Songs, *Qawwali*,
k) Poetic Recitations by Poets,
l) Obituary Programmes, and
m) Election Broadcasts.

The project for the transfer of sound archives on compact discs, was approved on 18 January 1993 at a cost of Rs.19.58 million, and was launched in 1999. Salient features of the project include installation of computerized sound lab recording equipment, conversion of analog magnetic tape recordings into digitalized archives, best possible mastering of sound quality and removal of noise, hiss and distortions from old recordings, and training of PBC staff. In the first instance PBC plans to transfer 50,000 minutes of recordings to compact discs over a period of two years.

23) Production of a series of 42 programmes under the title 'Awaz Khazana' based on archival material available with CPU for broadcast from Stations.
24) Preparation of 32 Long Playing Records, which are a pride production of CPU. These include three LPs containing 30 flood songs.
25) Setting up of a Music Instruments Museum. Music instruments of renowned vocalists and instrumentalists who are no longer alive, have been collected and are on display at CPU, Islamabad.
26) Recorded features, musical features and documentaries, etc. sent to ABU and Hoso Banka, Japan for participation in prize competitions. Some of the entries won prizes.
27) Special programmes on *Ramazan-ul-Mubarak, Muharram-ul-Haram, Rabi-ul-Awwal, Eid-ul-Fitr, Eid-ul-Azha*, anniversaries of the four Caliphs and others.
28) Special programmes for broadcast in the national hook-up as special election broadcasts since 1980. Catalogues also brought out on these special occasions.
29) Scripts of different programmes supplied to research scholars, National Book Foundation and Quaid-i-Azam University for preparation of research papers and books.
30) Special 70 talks in Urdu by Maulana Muhammad Umar under the main title 'Hikayat-i-Hikmat' for broadcast for moral uplift.
31) A set of two CDs, first ever of Radio Pakistan, titled '*Javedan Noor Jahan*' (Long Live Noor Jahan) based on selected items available in the archives. One CD contains 15 songs sung by Noor Jahan during the wars of 1965 and 1971 while the other CD has nine different songs, *ghazals* and poems in her voice.

NATIONAL SOUND ARCHIVES

The Central Library contains tape-recorded speeches of eminent persons, musical presentations by outstanding musicians, poetry recitals by renowned poets and documentaries on development projects and events of national importance. Some of the gems in this treasury of tapes are recordings of departed leaders, masters in the world of arts, men of letters and religious divines.

About a hundred countries have arrangements or cultural agreements with Radio Pakistan for the supply of material on tapes from its Central

16

Pakistan Broadcasting Academy

Staff Training School

In view of the great shortage of professionally trained radio producers, newsmen and technicians for running the programmes and handling the transmitters, the need for a training institute was felt at a very early stage. A senior Station Director, Sajjad Sarwar Niazi, who had himself had his training at BBC, was appointed the first Director of Staff Training in January 1949. He prepared a scheme for a Staff Training Centre that came into being on 7 November 1949. The staff was trained in the theory and practice of broadcasting to enable them to perform their duties more efficiently. The Centre was housed in the extension blocks of the new Broadcasting House then under construction at Karachi.

The first course for the preliminary training of technical probationers and Technical Assistants began on 11 November 1949 for five months. The second four-week course began the following month to provide elementary training to the newly recruited programme staff. The third and fourth courses began on 27 March 1950. The third course was for scriptwriters and was attended both by members of Radio Pakistan and outsiders interested in writing for radio. Inaugurating the course the Controller of Broadcasting said: 'Each station of Radio Pakistan consumes words at the rate of 80,000 words per day.' Bokhari further said: 'We want to train new radio writers not only in the technique of radio writing but also in the substance of our nation-building programmes viz. Islamic history and culture; the contribution of Muslims to literature, art, music and science; Pakistan and her problems and in subjects of general interest to our vast listening public.'

The fourth 10-week course was an advanced course for the programme staff. From 1 May 1953, the activities of the School were suspended owing to financial stringency.

TECHNICAL TRAINING SCHOOL

In 1961, a separate Technical Training School was established with I.A. Ansari as head to impart training to technical staff of Radio Pakistan. Later both the Schools were shifted to barracks of Pakistan Secretariat, Karachi, where they continued up to April 1970. In May 1970 both the schools were shifted to Islamabad, Sector H-9 in a newly constructed building, which was specially planned for the two schools along with a hostel.

MERGER INTO STAFF TRAINING SCHOOL/PBA

After the incorporation of Radio Pakistan the two Schools were merged in 1973 into one institution, the 'Staff Training School' as was the case during the first 12 years (1949-1961). The purpose of the merger was to ensure better coordination of the PBC training programmes. The school was redesignated as Pakistan Broadcasting Academy in 1983 and the news wing was added in 1985. Each Department is headed by a Vice-Principal and manned by a number of lecturers and demonstrators drawn from experienced officers of the Corporation. A Principal coordinates the training programmes. At the Headquarters, the Director of Administration coordinates the training with the recruitment policy.

With the elevation of status the Academy expanded its role in the area of training. A higher level of refresher courses is taken up at the Academy. Regional and sub-regional activities of the Asia-Pacific Institute for Broadcasting Development, Deutsche Welle of Germany, UNESCO, and Commonwealth Broadcasting Association, are also organized as and when they arrange foreign experts/resource persons for conducting different courses in the three job disciplines of broadcasting. The training arrangements and facilities at the Academy are of such a standard that AIBD had decided to hold its sub-regional courses on a regular basis. In these courses professional personnel from broadcasting organizations of Asia and the Pacific region as well as Oceania and Africa are trained. Courses in media management and production for producers are conducted in collaboration with AIBD and Deutsche Welle.

The Academy caters not only to domestic needs but has opened its training facilities to the broadcasting organizations of neighbouring and friendly countries like Iran, China, India, Sri Lanka, Maldives, Bangladesh, Bhutan, Nepal, Central Asian Republics and to trainees from the Middle Eastern and African countries. The Academy has also sent experts to other countries on the call of the Asia-Pacific Broadcasting Union or AIBD for imparting training in member countries. The Academy is thus heading towards meeting the domestic and regional needs in the field of broadcasting and keeping pace with the demands, within its limited resources.

Types of Training/Courses Available

Two main types of training courses are offered at the Academy—induction course for the new entrants and refresher and advanced courses for the in-service personnel. The induction courses are for apprentice producers and engineers, who hold either a Master's or an Engineering degree. At the PBA they receive a three-month preliminary training in programme production or broadcast engineering. The objectives of these courses are to give the trainees basic understanding and practice in their professional assignments in Pakistan Broadcasting Corporation. They are given simulation exercises in equipment handling, production and presentation of various types of programme to make them useful when posted at the operational units. The induction courses conclude with written examinations designed to test the concept, skill and practical knowledge of the trainees. Those who fail in the examination are laid off. The successful candidates are attached to different PBC units for another three months for on the job training. On the successful completion of the training, the apprentices are appointed as Producer or Technical Assistant.

The preliminary training given in the induction courses is not the end of the training process. Broadcast engineering, programme planning and production techniques have become too complex and sophisticated to be fully understood and mastered by novices in the short period of three months. Moreover, developments in broadcast equipment and production techniques are taking place so rapidly that what one learns today becomes outdated within a couple of years. Unless the PBC engineers and producers are kept abreast of these new developments and trained to use the new techniques they would become redundant. Moreover, due to the limited training facilities there is a backlog of engineers and producers working in PBC who have not had any training. They have to be trained and their skills updated. The Academy, therefore, arranges basic and advanced training courses for the in-service engineers and producers. The duration of these courses ranges from four weeks to twelve weeks.

The training courses available with the Academy include broadcast technology, broadcast engineering, semi-conductors, maintenance, operation and fabrication of transmitters and studio equipment, aerial systems, logic circuits and digital techniques, multi-band antenna system, modulation system, measurements and measuring equipment, design and acoustic treatment of studios, satellite communication, radio programme production (basic and advanced), interview techniques and investigative reporting, art of presentation, idea formulation for radio programmes, radio commentary, radio drama production, history of Pakistani music, radio for women, children and elderly people, programmes for rural women, development support broadcasting, OBs and radio reports, environment awareness programmes, planning and production of science programmes, social sector development and radio, farm management, rural development broadcasting, programme management, public service programmes, health promotion progrmme campaign, audience research, radio journalism, reporting and news editing, newscasting and actuality inserts, scriptwriting, news orientation, regional, local and external news production, broadcast management, training methodology, etc. Special courses can be mounted in broadcast engineering, programme production and radio journalism for a specific group of trainees on advance request from a broadcasting organization.

Facilities available at the Academy include a hostel with arrangements for 60 trainees at a time, lecture rooms fitted with film/ slide/overhead projectors, an auditorium, five sets of studios fully equipped, a FM transmitter of 10 watts and an AM transmitter of 1 KW, four communication receivers, two electronic laboratories (basic and digital), modern measuring kits and equipment, and a library with adequate collection of general and technical books and periodicals. There is a Central Laboratory and Workshop on the PBA premises where major repair and renovation of all types of electrical and electronic equipment is carried out on centralized basis. Although it is not an integral part of the Academy, nevertheless it provides a convenient technical area for practical training in repair and maintenance of broadcast equipment, transformer winding, workshop practices, etc.

On the Engineering side all departmental candidates for the post of Sub-Engineer and Broadcast Engineer are given three months training followed by an examination. Their promotion depends on their success in the examination. This is a good practice and needs to be emulated by other departments and its scope widened to include higher posts as well. During the training period, guest lecturers, who are specialists in their field of study, are invited from within and outside the Corporation and from Universities. Lectures are usually preceded by cyclostyled synopsis and followed by group discussions and seminars. In the case of induction courses for new programme entrants, lectures to explain the various stages of the process of the transmission of sound are delivered for an elementary understanding of the technical aspects of radio communication. The working of the equipment that makes radio communication possible is also demonstrated.

During 1975-76 a six-week workshop on Multimedia Approach to Learning and Educational Broadcasting was organized in collaboration with the Open University and the British Council. It was attended by 14 Producers/Senior Producers of PBC and 20 teachers from the Open University. In the same year a 4-week course on Rural Broadcasting was organized in collaboration with the Integrated Rural Development Programme. It was conducted by an expert from BBC and attended by 10 Producers from PBC and four Agricultural Officers from the Provincial Departments of Agriculture.

Since conversion into a Corporation, the Pakistan Boradcasting Academy has conducted 158 courses. Of these 65 related to Programmes, 69 to Engineering, 21 to News and 3 to Training Methodology. During the same period a total of 2002 trainees have gone through these courses. Their composition was 784 from the Programme, 970 from Engineering and 218 from the News Wings and 30 in Training Methodology. These included 57 trainees from foreign countries. Of the total courses 30 were held in collaboration with foreign broadcasting organizations/bodies: AIBD (19 courses), SAARC (3 courses), Deutsche Welle (2, in one of these Goethe Institute also cooperated), CBA, USIS and UNICEF collaborated in one course each. In these courses 325 persons received training.

EXPERT ADVISORY SERVICES

PBC also provided expert services to friendly countries under the aegis mainly of AIBD and ABU. Only three persons have gone abroad on this account. They are Zamir Siddiqui, Vice Principal, to Malaysia, Papua New Guinea, Fiji and Nepal, Younus Khan, Principal, to Papua New Guinea and Malaysia under AIBD and Malik Aziz Khan, Vice Principal, to Kiribati under the Asia-Pacific Broadcasting Union.

17

Audience Research

Radio is regarded as 'the people's medium' and in order to retain its popularity in the face of increasing competition from other media sources, broadcasting authorities must stay in tune with public taste and preferences and keep informed of changes in listeners' choice through feedback. No broadcasting organization can afford to ignore the vital need for proper audience research.

Functions

The function of audience research is to get listener feedback on radio programmes, to analyse the data so collected and prepare research reports for action by programme planners and producers. It also aims to prepare a listeners' profile in respect to demographic characteristics like residence (rural/urban), age, sex, education, likes and dislikes, tastes and preferences. Audience researchers suggest ways and means to modify programme policies in the light of listeners' reactions and suggest improvements in current programmes, and collect fresh ideas for new programmes. It assesses programme policies affecting the quality and standard of programmes.

Audience research can also determine whether Radio Pakistan is losing its listeners to foreign broadcasting organizations, what programmes are acceptable or popular, what the majority of listeners seek from Radio Pakistan and whether this is provided. Researchers maintain liaison and exchange material on audience research with other broadcasting organizations, develop statistical analyses of audience research, convene conferences and meetings relating to audience research, maintain statistics on the composition of programmes and listener feedback of various stations.

Background and Present Position

In Pakistan, audience research began with the appointment of Listener Research Officers (LROs) at Karachi and Dhaka in 1951. The first LRO began work on 10 May 1948 but he resigned after a short period. The Muslim LRO in AIR, who was a qualified sugar technologist, on coming over to Pakistan was posted as Assistant Station Director, Peshawar, and retired as Joint Director General, Radio Pakistan. In West Pakistan, the post was held to September 1955. After a lapse of about five years one LRO was appointed at Dhaka and another at Karachi in 1960. During the nine-year period from 1960 to 1969 the LRO at Karachi conducted 44 surveys.

Audience research was regularized with the appointment of a Director, Listener Research at the Directorate General on 1 June 1965 and LROs during 1966-67 at seven more stations bringing the total number of LROs to nine—six at Karachi, Lahore, Quetta, Peshawar, Hyderabad and Rawalpindi in West Pakistan and three at Dhaka, Chittagong and Rajshahi in East Pakistan. From this date to 1971 useful audience research was a great benefit to programme planners. Even during the year 1971 when the unsettled political situation hampered work, as many as 23 surveys were taken up, 18 in West Pakistan and five in East Pakistan. These surveys garnered feedback on all radio programmes from talk shows to entertainment and educational broadcasts as well. Special reference to new programming and transmission times was incorporated in the surveys to get listener response.

Thereafter, the audience research set-up at Stations was allowed to liquidate itself. It was expected that with the conversion of Radio Pakistan into a statutory Corporation, audience research would get the importance it deserved. A change was perceptible in the outlook, attitude and emphasis. It, however, took the Corporation seven years to re-start audience research towards the end of 1979. However, a small unit of five persons (one Audience Research Officer, one Investigator and three Research Assistants) was set up at Peshawar in 1974 for specific purposes.

The composition of the present set-up is one Controller/Deputy Controller, two Audience Research Managers, eleven Audience Research Officers and nine Research Assistants. Since its reorganization in 1979 till 30 June 2000 nearly 143 audience research studies have been conducted. Almost all the studies are local or regional in nature, and cover a variety of subjects and programmes for different segments of society.

Surveys conducted on nation-wide basis related to programmes on current affairs, religion, entertainment, listening pattern and those meant for the rural population, farmers, women and the youth. The nation-wide survey on programmes for the youth coincided with the International Youth Year. A nation-wide study on SAVE programmes was also undertaken. These SAARC programmes are broadcast on exchange basis among the broadcasting organizations in member countries every month. Another important survey was on Iodised Salt and Goitre Disease in Northern Areas, Azad Kashmir and Rawalpindi Division conducted on behalf of the Ministry of Planning and Development, Islamabad.

Multi-stage sampling technique is used for the selection of urban enumeration blocks and villages. The systematic sampling method is used for the selection of households and respondents. Data is collected on a pre-designed questionnaire through interviews of the sample audience at their homes or place of work. On a few occasions the group discussion and panel methods have also been employed. In the fifties and sixties and also in the early seventies the panel method was generally used. Questionnaires were mailed to selected listeners and panel members and reports were prepared on the basis of the responses.

A new dimension was added to audience research at PBC to determine the growth or need survey conducted before starting a new station or service. These surveys provided useful and vital information for planning programmes for the new station or service to meet the preferences and expectations of the people of the target area.

Other ways of getting listeners' opinions were through mail and telephone calls received at stations, deliberations of advisory committees, news items, comments, reviews, columns and letters published in national and local newspapers and journals and sometimes parliamentary discussions. But public opinion thus obtained neither constitutes audience research nor is a substitute for it, because they are not the result of a systematic study where different variables can be controlled.

Letters and telephone calls are received in thousands every month. They are voluntary and spontaneous and are generally received from the more vocal elements of society who are more interested in hearing their names on the radio and who generally belong to comparatively younger age groups. Press comments represent the view of the reviewer or correspondent. The majority of letters received by stations are requests for songs. Few letters contain suggestions and comments, critical or appreciative. Such letters are replied to in a weekly programme broadcast locally from all stations. Letters addressed to specific programmes are acknowledged in those programmes. Still these letters are studied and tabulated at each station and a report is sent to the Headquarters every month. In the early days when audience research did not exist, comments contained in listeners' mail and comments appearing in the press were taken into consideration whenever a new programme policy was undertaken.

Gallup Survey

In 1999, for the first time PBC had a sample survey taken by an independent agency, Gallup Pakistan. The findings were mainly for the consumption of advertising agencies and advertisers and also as a measure to check the veracity of its own research studies. The sample study polled 7,000 urban and rural adults of 18 years and above. The findings made interesting reading.

According to the survey, seven million Pakistani adults in the rural areas and three million in the urban areas listen to the radio. This amounts to almost one-third (34 per cent) o the adult population of the country. Not all of them are regular listeners. Some tune in radio sets once a month or on a special occasion like a crisis situation, special event or leisure past time. Of this listenership, 22 per cent listen to the radio regularly—males having an edge over females. Listening decreases with increase in age but rises with increase in income and education.

The survey showed that music programmes are the most popular followed by news and religious programmes. Prime listening timings are around 0700 hrs, 1100 hrs and 2000 hrs. FM channels are popular in the three cities where they are located—Islamabad/Rawalpindi, Lahore and Karachi. About 1.5 million adult men and women are frequent listeners of FM radio in these cities, even the late hour programmes. Reasons given for not listening to the radio included not having a radio, no interest and no time.

So far two Conferences of Audience Research Officers have been held. The first was in Karachi in April 1966 and the other in June 1972 at the Headquarters in Rawalpindi. Useful and workable recommendations were made for improving and reorganizing audience research set-ups in the proposed Corporation. Some of these are being implemented.

Some foreign broadcasting organizations periodically carry out audience surveys to determine interest in their external broadcasts directed at Pakistan. These organizations include BBC, Voice of America, Radio Beijing and Radio Japan. The USIS undertook a study in 1958-59 to obtain a profile analysis of radio listening habits and preferences among those Pakistanis who were registered radio owners as determined by the radio licence file with the postal authorities. It may be of interest to know that in Japan a nation-wide 'census' of radio listeners was done in May-August 1932.

The Mumtaz Hasan Broadcasting Committee recommended the setting up of a well-staffed centralized Audience Research Cell for both Radio and TV. It also recommended that Audience Research Officers should be appointed at regional stations.

18

PRESS, PUBLIC RELATIONS, PUBLICATIONS AND OVERSEAS LIAISON

PRESS AND PUBLIC RELATIONS

Radio Pakistan had a Director of Liaison, whose functions were split after incorporation. One set of responsibilities relates to press and public relations and the other to publications. The Corporation publishes annual reports, house journals, company literature and greeting cards. The Head or Deputy Controller of Press and Public Relations is responsible for press liaison and coordination with the Government.

Newspapers and periodicals are scrutinized daily and press clippings kept for the information of the Director General and other members of the Board of Directors. Whenever necessary, clarifications are issued to the press. Press releases on important and topical national hook-up items are sent to the media regularly. Publicity is given to programmes for special events, quarterly review of programmes, PBC news-making events, and the President's and Prime Minister's engagements. All stations and the External Services have regular programmes to respond to the large number of letters received from listeners from within and outside the country. Dxers' SINPO reports are suitably acknowledged and verified through QSL cards. The Section conducts tours of Radio Pakistan for distinguished visitors and delegations, both local and foreign.

The Section published PBC Yearbook for the years 1974-75 and 1975-76; and for 1973-74, 1976-77 and 1977-78 in mimeograph form. During 1974-75, it brought out a house journal *'Khabarnama'* for distribution within the company to improve employee-management relations. Only a few issues were published.

Publications

The Indian Broadcasting Company published the first issue of 'The Indian Radio Times', the fortnightly programme journal on 16 July 1927 containing interesting reading material, advertisements and programme details of the Bombay Station that went on the air on 23 July 1927. It was shifted to Delhi in August 1937. Coinciding with the opening of Delhi Station of the Indian State Broadcasting Service on 1 January 1936, the journal came out in a new format and under a new name 'The Indian Listener'. It contained programme details for the first fortnight of January 1936 of Bombay, Delhi and Calcutta Stations as well as of BBC. *'Betar Jagat'* the programme journal in Bengali started in September 1929 from Calcutta. The Hindustani programme journal *'Awaz'* in Urdu and Hindi, started publication from Delhi from 1 January 1936, the day of inauguration of the Delhi Station. It was bifurcated into *'Awaz'* in Urdu and *'Sarang'* in Hindi in July 1938.

Ahang and Pakistan Calling

In Pakistan the first issue of *Ahang* was out on 14 August 1948, coinciding with the anniversary of Independence Day and commissioning of the Karachi Station. The 40 pages of the journal contained editorial, photographs, reading material and programme details for 14-31 August 1948 of Karachi, Lahore and Peshawar Stations and selected programmes of Dhaka Station. The insignia on the cover page was by the celebrated painter. The first issue of the fortnightly English programme journal 'Pakistan Calling' came a few days later on 25 August 1948. The two illustrated journals contained programme details, photographs and ensembles of the programmes and events covered, extracts from previous broadcasts and advance notes on important programmes scheduled to come on the air. Both the journals were for subscription and were run on a commercial basis.

Full details of programmes of Dhaka Station were published in 'Pakistan Calling' from the second fortnight of January 1949 synchronizing with the installation of the first short wave transmitter of 7.5 kilowatts in Dhaka in January, 1949. From May 1950 the size of the two journals was increased. In early fifties, *Ahang* carried programme details of Azad Kashmir Radio as well. From January 1951, 'Ahang' reserved one or two pages for children and one page for photographs. 'Pakistan Calling' was re-designed from 1 January 1951 and more attention was paid to the magazine section so that it became a radio-cum-cultural magazine. This continued up to July 1952 when its size was reduced to that of a programme bulletin.

'Elan' was the programme journal in Bengali and was published from Dhaka. It started publication from 1 August 1951 and contained programme details for 14-31 August, 1951. As sales floundered it was converted into a programme bulletin from 1 April 1954 on the pattern of 'Pakistan Calling'.

Another programme pamphlet 'Azaat-i-Pakistan' began publication with effect from 1 August 1950. It contained programme details of both the chunks of Arabic Service with suitable illustrations. Two more programme pamphlets, 'Sada-i-Pakistan' and 'Pakistan Athan' were published in Persian and Myanmar (then Burmese) for programmes of the Persian and Burmese Services on the pattern of 'Azaat-i-Pakistan'. The Arabic and Persian programme bulletins were published in Karachi, while for the Burmese programme bulletin, the dummy was prepared at the External Services Karachi and sent to the Pakistan Mission in Rangoon for publication and distribution. All the three programme bulletins were distributed free among interested listeners in the target areas. These were later discontinued and programme details, etc. of all the External Services are published regularly in the monthly programme bulletin 'Pakistan Calling'.

In 1953, due to cuts in the budget grant of Radio Pakistan the publication and distribution of 'Ahang' and 'Pakistan Calling' was entrusted to a private publisher on a no-cost basis for the Government. The policy of the journals was, however, controlled by Radio Pakistan.

During the year 1963-64, the printing of 'Ahang' and 'Pakistan Calling' was taken over by Radio Pakistan from the private publishers and the price of *Ahang* was reduced to boost its sale among the average income groups. Its design was also improved.

Due to the September 1965 war, all normal programme schedules were changed and it was not possible to publish details in 'Ahang'. The pages of the journal were reduced from 72 to 48, displaying only a bare skeleton of programmes. This adversely affected its circulation that came down from 5,000 to 2,000 copies only. Many of the sale agents cancelled their agencies or reduced their demand. Despite unfavourable circumstances a circulation of 2400 was maintained in 1966 also. 'Pakistan Calling' was brought out in the form of a pamphlet.

Soon after incorporation of Radio Pakistan in December 1972 the fortnightly 'Ahang' was re-designed with the result it gained popularity and its circulation increased rapidly. Its design, size and format were changed, and it competed favourably with the commercially-run weekly and fortnightly social, cultural and literary magazines. Its sale exceeded 12,000 copies and the firm demand for the next issues was much higher. It became one of the leading Urdu journals of the country. Its subscription list included fans from Hong Kong, Malaysia, Gulf States, Iran, Kenya, Denmark, Norway and the United Kingdom. The journal's popularity was achieved by making its appearance more attractive and its contents more interesting.

The regular features in *Ahang* included *Raushni*, scripts of literary pieces broadcast by various stations of PBC and interviews of radio personalities. Classic novels, both local and foreign, were serialized. Effective from January 1976 *Ahang* became a member of the All Pakistan Newspaper Society. The golden period of popularity for *Ahang* continued till 1977. Thereafter its format and contents were changed reducing it to the level of an official or house journal, with the result that sales dropped drastically. It is being published in the same manner since then with minor changes.

Since March 1987, *Ahang* has become a monthly magazine printed on art paper with the number of pages doubled to 112. The first half of the magazine contains a selection of informative and educative articles and poems already broadcast by PBC stations/units. The second half of the magazine comprises details of anniversary and festival programmes, timings of the External Services and the World Service, transmission timings and frequencies of the programmes of PBC stations, highlights of religious programmes, national and provincial hook-up programmes, fixed-point chart, schedule of all national, regional and local news bulletins, salient features of the special audience programmes and 'Producers Themes'. The last item has since been discontinued.

At the present time *Ahang* has 76 pages and is the only magazine of PBC with a literary touch. It devotes one or two pages for children's features. During 1993, it was decided to publish special *Ahang* issues for the inauguration anniversaries of various stations. The first issue was published in August 1993 for Hyderabad. It may be of interest to know that a student of Jamshoro University in the Master's Degree programme in Mass Communication wrote a thesis on *Ahang* that was approved in 1991.

'Pakistan Calling' was also originally a literary magazine as it contained specially written articles, broadcast material, photographs, news and events about Radio Pakistan and poems besides programme details of both Home and External Services and Western music. From the beginning of 1963, it is being published as a monthly magazine. During the sixties and seventies, it mostly carried transcripts of talks broadcast by VVIPs or by Heads of Ministries/Departments about progress made by their organizations, current affairs items and details of short wave frequencies and External Services. It took the form of a pamphlet. The number of pages was reduced further and for some time it had only eight pages containing timings, frequencies, titles of programmes of External Services and World Service in English, Urdu, Arabic and Iranian. At one time it took the form of a folder and a leaflet. Now, it has twelve pages with an attractive cover, photographs, highlights of programme schedules, talks broadcast from External Service about various aspects of Pakistan, and programme details in English, Urdu, Arabic, Farsi, Turki and Dari. It is distributed free to listeners in the target areas either direct or through Pakistan Missions abroad.

The Publication Section at Karachi is headed by a Deputy Controller and assisted by an Editor (Programme Manager), two Sub-Editors and other supporting staff.

Important Publications

Some important publications of the Corporation besides the monthlies *Ahang* and 'Pakistan Calling' are:

i) *Zindabad*—collection of broadcast talks, poems, features by leaders, poets, writers paying homage to the *Quaid-i-Azam*, 25 December 1950.
ii) Three Years of Radio Pakistan, 1950, Karachi.
iii) Handbook for Junior Programme Staff, 1951, Karachi.
iv) Twenty Years of Radio Pakistan.
v) Special numbers of 'Pakistan Calling' and *Ahang* on the occasion of Development Decade 1958-68, Centenaries of *Quaid-i-Azam* and Allama Iqbal, 1965 and 1971 war Silver Jubilee of Radio Pakistan, etc.
vi) Ten Years of Development, Radio Pakistan: 1958-68.

vii) 'Quran-i-Hakeem aur Hamari Zindagi' Vol. I and II—collection of talks broadcast under the same title on Islamic teachings and their application to daily life.

viii) 'Raushni', Vol. I & II—collection of 87 and 90 talks respectively broadcast under the same title on moral and ethical aspects of human life in the light of Quran and *Sunnah*.

ix) 'Islami Kahanian'—a selection of 34 stories for children broadcast from Radio Pakistan.

x) A Tragedy in Focus—a series of 7 talks broadcast in External Services of Radio Pakistan—a narrative in human terms of what really happened in East Pakistan, September 1971, Karachi.

xi) 'Nai Tarz-i-Nashriyat' (New Concept of Broadcasting), Vol. I (1987) and II (1988-91).

xii) Booklets on *'Islamiat'* and *'Musalman Mosiqar'* based on broadcast material brought out by the Department of Advertising, Films and Publications, Government of Pakistan in early fifties.

Books at S. No: vii, viii and ix are available with Radio Publications, Pakistan Broadcasting Corporation, Broadcasting House, M.A. Jinnah Road, Karachi.

The Mumtaz Hasan Broadcasting Committee recommended the formation of a powerful and well-staffed centralized public relations and publications cell for both Radio and Television. It also stated that Public Relations Officers should be appointed at regional stations.

OVERSEAS LIAISON AND TRAINING

In Radio Pakistan the function of Overseas Liaison and Training was also looked after by the Director of Liaison. In PBC this is the responsibility of a Deputy Controller, known as H (TOL) Head, Training and Overseas Liaison.

Membership of International Broadcasting Bodies

Radio Pakistan is the founder, full member or associate member of a number of international broadcasting and related bodies such as ABU, CBA, ISBO, ASBO, AIBD, EBU, International Folk Music Council, International Institute of Communication and International Advertising

Association (IAA), etc. Radio Pakistan has been taking active part in the deliberations of these organizations except in cases when a delegation could not go due to some natural calamity, emergency or any other pressing problem necessitating presence in the country. It participates in the conferences, seminars, symposia, prize competitions, music festivals organized by these organizations as well as those of ITU, CCIR, IFRB, AMIC, ECO, OIC, etc.

Besides general cultural exchange programmes between countries, Radio Pakistan has bilateral agreements with broadcasting organizations in about 60 countries. Under these general cultural exchange programmes and bilateral agreements broadcast material is exchanged and PBC officials go abroad on study tours or goodwill exchange visits.

FOREIGN TOURS/VISITS

During the period from 1990 to June 2000 about 112 officers went abroad for the coverage of the President's and Prime Minister's visits to foreign countries. During the same period 102 PBC officers went abroad to attend seminars, meetings, workshops and conferences, etc.

PBC's sports programmes are highly popular with listeners. In addition to sports events at home, all international sports events like Olympic and Asian Games, world championships, and international hockey and cricket matches are covered via satellite. During the period from 1981 to June 2000 as many as 249 producers and commentators went abroad for the coverage of international hockey and cricket events.

PBC officials are sent to various countries of the world particularly to Malaysia which houses the training centre of AIBD in Kuala Lumpur. During the post-Corporation period from 1973 to June 2000 about 199 PBC officials were deputed for specialized training in the field of radio programme production, engineering, radio journalism, audience research and management to countries like Australia, Japan, Malaysia, Germany, Netherlands, UK, USA and Canada.

PBC officials also receive training in different institutions within the country in such fields as administration, management, trade unionism, systems analysis and civil defence. The institutions whose services are availed of include: National Institute of Public Administration Karachi, Civil Defence Academy Lahore, Secretariat

Training Institute Rawalpindi and Pakistan Institute of Management Karachi.

AIBD Governing Council Meeting

PBC and PTV hosted the 25 annual meeting of AIBD Governing Council, held at Islamabad from 19 to 22 July 1999. About 35 delegates from Malaysia, Indonesia, Thailand, Singapore, Bangladesh, China, Philippines, Iran, Nepal, Sri Lanka and India attended. Secretary General of ABU, Project Director of Friedrich Ebert Stiftung, Germany and Executive Director of Radio Canada International also participated. The Executive Committee of AIBD met on 19 July. The Information Minister formally inaugurated the Governing Council meeting. The Director General, PBC, was elected Chairman, Strategic Plan Team. Its main purpose was to devise/define the sphere of activity and strategy for the Government controlled broadcasting organizations. It also recommended use of the latest information technology facilities including the Internet.

19

PAKISTAN BROADCASTING FOUNDATION

FUNCTIONS

In January 2000, PBC established a Foundation under the style of Pakistan Broadcasting Foundation (Guarantee) Limited incorporated under Section 43 of the Companies Ordinance, 1984. The Foundation is to work solely for the welfare of PBC employees and their dependents. The underlying idea of PBF is to handle income-generating projects that PBC as a broadcasting organization is not in a position to undertake.

Briefly the projects that the Foundation can undertake, promote or support include: audio text and other telecommunication services, advertising, publishing, film-making, drama production; designing and manufacturing of equipment, systems and services for computers, telephones, radio, voice-paging and other forms of communication and telecommunication; establishing and running schools, academies and other educational and welfare institutions including hospitals, clubs, etc.; to promote art, science and culture; support the welfare of current and former employees of the Company and PBC holding seminars, workshops, exhibitions, etc.; working as civil, mechanical, sanitary, watering, plastic and metallurgy engineers, and as architects, consultants, etc.; undertaking all types of construction works including buildings, roads, bridges, tube-wells and the like.

Some of the major works undertaken by the Foundation are described below:

BOYS COLLEGE: The first project of the Foundation was a Boys College established in April 2000, to provide quality education to the children of PBC/PTV employees at relatively affordable cost. The college was established in collaboration with the OPF Girls College on the premises of the Pakistan Broadcasting Academy, Sector H-9, Islamabad.

The first academic session of the college began on 19 April 2000, with a total of 324 students. The children of PBC/PTV employees admitted at this college were allowed 50 per cent concession in tuition fee and free enrolment in sports activities. They were also exempted from payment of security deposits. The college comprises a modern complex with fully furnished computer labs, broadcasting studios, air-conditioned cafeteria and sports facilities including swimming, tennis, baseball, cricket, badminton, etc.

HOUSING PROJECT: PBC has 39 flats in Sector G-8/4, Islamabad, allotted by CDA in 1998. It was decided to establish a Housing Foundation for PBC employees. The Foundation had to undertake the construction of flats on vast unused PBC land in Islamabad, Lahore and Karachi in collaboration with Messrs National Construction Ltd. while Messrs Nespak were to be the Consultants. A total of 1618 flats were to be constructed for allotment to PBC employees on an ownership basis. The modalities of the project were briskly underway when the government in power started the '*Mera Ghar Scheme*' and PBC land in several cities was taken over by the Prime Minister Housing Authority.

The National Security Council and the Cabinet in a meeting held on 9 March 2000 approved implementation of the housing project of PBC at Islamabad on the land situated in Sector G-8/4, Islamabad. PBC has taken up the project of constructing 440 flats of category II, III, IV, V and VI in Sector G-8/4, Islamabad. The exact number of apartments is to be finalized after the approval of the plan by CDA.

VOICE-PAGING: PBC has plans to launch a voice-paging system in Pakistan to augment its revenues. For this purpose it obtained the necessary licence from PTCL to operate the system and signed agreements with two private parties. One agreement is to start the service as a joint venture. PBC has no financial obligation as the entire initial cost is to be borne by the private party. This covers purchasing equipment, voice-page sets, establishing and operating the service. PBC will allow use of idle side bands of FM 101 as sub-carrier for voice messaging. Profit sharing between PBC and the private party is to be in the ratio of 62:38. PBC will also get Rs.200 per new subscriber as activation premium and its mono FM transmitter will be converted into stereo with no cost to PBC.

Subscribers can receive weather updates, headline news, stock exchange quotations, sports updates, Radio and TV programme schedules, flight and train timings and alarm and *'Azan'* services at a very nominal cost. The second agreement signed with another private party is regarding marketing rights of FM 101.

The service is to be initially launched in Islamabad and soon thereafter made available in Karachi, Lahore and Peshawar. It is to be expanded to Faisalabad, Sialkot, Gujranwala, Hyderabad, Sukkur, Multan, Jhelum and Quetta. In fact it will cover all the areas of Pakistan where FM 101 transmissions are available.

20

The Future

Potential of Radio

Radio in Pakistan has unique advantages not matched by any other medium, print or electronic. Its transmission hours run through the day to the early hours of the morning. Broadcasts are made in most of the languages and dialects in the country and hence Radio's vast appeal with the listening population. Programmes can cater to different tastes, age groups, literacy levels and specific needs. An important factor is that it addresses those segments of society that can influence a behavioural change in society.

Radio Pakistan is committed to playing an effective role in any Government project aimed at the welfare of the people. One way of doing this is to influence a change for the better in society. Through its programmes it has the power to break down prejudices and negative thinking, encourage problem-solving and self-help, educate through dissemination of information, and support the activities of the government in social uplift.

To do this, Radio needs the assistance of policy makers, communicators, subject specialists and above all those who allocate vital resources.

It is unfortunate that Radio's potential has not been adequately exploited. It still remains a greatly under-used medium in Pakistan although it is very cost-effective and influential. Absence or shortage of electric power is no obstacle to radio accessibility since the advent of the transistor and integrated circuits. This is a significant factor in Pakistan as nearly 70 per cent of the population lives in rural areas and barely 35 per cent of the villages have access to electricity. Radio Pakistan broadcasts in as many as 20 regional languages in addition to the national language. This provides radio with a unique ability to speak to the target audiences in their own tongue and idiom. To exclude radio from any communication strategy is, therefore, a serious flaw in national planning.

The range of radio programmes is as vast and varied as this diverse country. Radio Pakistan, being the largest and universally accessible mass medium in the country, has the manifold task of informing, educating, and entertaining people on a myriad of issues from ideology to the economy to culture, and ultimately anything that touches their lives. In short, the entire gamut of programme activity retains the basic underlying objective, i.e. to instil patriotism for their homeland, to inform of the world around them, to educate and motivate them to strive for a better tomorrow.

CHALLENGES

Each day brings new challenges to national life and in radio these challenges need to be met through new programme ideas, new production techniques, new methodology and new dimensions. On the one hand latest information is to be imparted on a much wider range of topics and on a much bigger canvas than ever before. On the other hand the listeners, through better awareness, are becoming more demanding with regard to the quality of programmes, and are setting new standards of accountability.

Radio Pakistan has faced another type of challenge several times. Whenever there is a revolution or counter-revolution, radio stations are among the first institutions that the new rulers take over in an attempt to validate themselves. The desire to use radio for party politics and even personal objectives has always been an unfortunate tendency in Pakistan. Radio has thus often been the target of political pressure and criticism from vested interest groups—political, linguistic, religious, cultural, etc. It is considered part of the 'game', or rather an unpleasant occupational hazard. In such situations the programme staff has to be assured of the understanding and support of the broadcasting authority or else the discharge of professional duties becomes difficult.

Radio today faces competition that was unheard of half a century ago, namely television and cable, video games and movies, computer games and the Internet. In the face of these alternatives, radio has to be very creative and imaginative to work in the sound medium alone and continue to capture its audiences. Radio faces tougher challenges than ever before, but for those who are drawn to it the personal and financial rewards are great.

Ways to Meet Challenges

PBC has another potential challenge from private broadcasters who may be granted permission to operate. In fact the challenge is already there, though on a limited scale, in FM broadcasting and in three cities only. The need for guidelines to protect against misuse of responsibilities by private broadcasters is urgent. It is understood that the Asia-Pacific Broadcasting Union (ABU) due to the growing emergence of private broadcasters in the region has been grappling with the problem of resolving differences between the two types of broadcasters—public and private.

Multi-channel or at least two-channel broadcasting is one of the answers to this situation for the simple reason that in every society there are conflicting demands on radio. Not all persons have similar likes and dislikes, tastes and preferences. Some would like to have more entertainment, others more information or educative programmes. It is not impossible to meet the demands of different categories of listeners provided adequate infrastructure and air hours are available. At present PBC is broadcasting the bulk of its programmes on a single channel. Even where another channel is available it is not being fully utilized because of the shortage of funds and lack of additional production facilities. Consequently PBC is unable to cater varied programme fare simultaneously.

To meet the imminent danger of private broadcasters and to improve the quality of its service, Radio Pakistan needs to expand its sound broadcasting so that the listener is able to make choices of alternative programmes in sports, music, news, religion, etc.

It is, however, absolutely necessary that people are given a choice to select any one of the programmes offered by the Home Service simultaneously. If they do not have this facility they may easily switch to broadcasts from other countries, some of whom are hostile to Pakistan. With the growing popularity of satellites, it is impertive to offer more programme variety simultaneously. For the time being only five PBC stations have the facility of a second channel. This facility has to be expanded to other stations. FM can be used as a third channel. Nearby overlapping stations can adjust their programmes with the same purpose in view. This will help in retaining audiences as well as attracting new listeners. It may also help in weaning away listeners of foreign broadcasts.

PBC's external services are radiated on short waves except for the Hazargi and Dari Services. Due to the popularity and ease of medium wave listening and its natural advantage over short wave broadcasts it is necessary to utilize medium wave transmission. A beginning can be made by broadcasting some of the external services directed to neighbouring countries on medium wave. Thereafter PBC can explore the possibility of setting up its own medium wave transmitters or hiring some outside the country to reach distant places on the globe as is being done by a number of leading countries. It may be mentioned that many broadcasting organizations in the world are already radiating their external services on medium wave transmitters. PBC would do well to follow this trend.

TECHNOLOGICAL DEVELOPMENT

In the fifties, radio ruled supreme as the sole medium for entertainment and information. After the seventies radio became the medium of choice in the kitchen and car in industrialized countries. Television invaded society and 'development support communication' became a common phrase. Broadcasting moved to another dimension, i.e. to inform, influence and motivate. Broadcasting became a tool for economic development and nation-building. In the nineties and in the early twenty-first century, broadcasting, in the words of Jaffar Kamin of Malaysia, had been transformed into a new realm of universal socio-economic and political interactivity to take a new dimension i.e. to inform, to interact and to integrate.

In the digital world of tomorrow, information technology will play an increasingly dominant role especially in the area of entertainment software. Information technology presents extraordinary challenges and opportunities for the developing world, which when adopted can lead to the success of countries like Singapore, Malaysia and Korea. These countries have comprehensive information technology and software development policies. Information technology is, therefore, playing an increasingly dominant role in the development of these countries.

The technological advances in communications have changed the quality of life and universalized human perceptions of global development. As a result of the unprecedented technological development in information technology 'one would be able to reach any library around the globe, request the playback of the latest film produced in Hollywood, send messages to friends and colleagues around the world and exchange information with others having similar interests. All the major newspapers and magazines around the world will be available for you to read and you don't have to leave home to have all this'. Even the office can be brought home.

This easy access to information has its disadvantages especially for nations seeking to protect their values and cultures. There is no distinction between a digital signal for an uplifting programme and that of a pornographic programme. Nobody can detect a bad signal. Pornographic images from beyond the border are receivable on home PCs and accessible to the young and adults alike. One solution is to prepare the future generation in how to deal with the bombardment of information, some of which is unwholesome. They must learn to be critical and responsible users of information, and to make intelligent judgements on what they see in the media. To achieve this objective each nation has to formulate its own system for a balance between freedom and responsibility of broadcast media.

Currently broadcasting services reach the audiences via four transmission technologies: (i) Terrestrial Broadcasting (the traditional method), (ii) Cable Networks, (iii) Satellite Broadcasting, and (iv) Microwave Multipoint Distribution Systems (MMDS). Radio Pakistan continues to make use of the traditional method. According to press reports one individual or group was granted MMDS service licence to operate radio, TV and telephone services in the country in April 1995. PBC is far behind other Asian countries such as Singapore, Malaysia, Indonesia and Thailand in the field of broadcasting hardware and software.

PBC's entire emphasis is on AM broadcasting when some countries have started doing away with AM. Although FM has been in Radio Pakistan from the very early days, it was used for the purpose of studio-transmitter link only and/or duplicating the AM broadcasts. It is only recently that PBC started using it for broadcasting programmes. Here also a better substitute is available in the form of digital audio broadcasting (DAB) which has been successfully experimented in some countries since the late eighties and early nineties and has been adopted by a few countries.

Compact disc (CD) was not used in PBC till mid-nineties when its replacement in the form of digital audio tape (DAT) was heralded in 1987. Transmitters inside and outside the Broadcasting House premises are operated manually in PBC whereas these have been remote-controlled in many countries for the last three decades or more. CAR (Computer Aided Radio) is yet another landmark in broadcasting. It makes big studios almost obsolete and gives radio producers independent means of producing programmes without involving other staff with whom they had to coordinate previously.

PBC's backwardness in the field of broadcasting can prove a blessing in disguise in the sense that it should go for the latest broadcasting technology and broadcasting hardware and software. For example, instead of transferring sound archival material to compact disc, PBC may go straight for digital compact cassette, or digital audio tape or at least digital laser disc. Similarly, instead of setting up FM network now PBC may skip over direct to digital audio broadcasting.

In view of the expensive and time-consuming option of replacement of old and installation of new AM transmitters, especially short wave, DAB offers the best alternative. This system is comparatively cheap in cost of equipment, offers variety of programme choices and wider radio coverage, gives superior quality of programmes and requires negligible maintenance and operational costs. Its adoption as full time broadcasting was delayed due to the absence of a digital receiver, which has since been developed. However, as the generation and management of the digital radio signal in the transmission chain is very different from that of a single channel AM or FM broadcast, it raises certain issues, mainly technical, which have to be tackled.

Problems and Solutions

Project planning in PBC has always been considered confidential, without consulting the users, with the result that a project is seldom successful. In most Western countries project planning involves almost all groups, parties and institutions who have an interest at any stage in the project. This is in addition to the actual users and beneficiaries of the project. In PBC the users normally are the programme developers and news personnel. They come to know of a project only when it is complete, i.e. when a transmitter has been installed or a Broadcasting

House has been constructed. The project plan has no input regarding provision for broadcast of programmes and news and consequently its requirements in men, material and money. There is practically no concept of research and development, without which no actual progress is possible. Research (including audience research) and training must be given due priority as they are basic requirements and provide the *raison d'etre* for broadcasting.

Some questionable decisions were taken at the time of converting Radio Pakistan into a statutory Corporation from a Government Department. These decisions were of very far-reaching nature and had a very debilitating effect on the objectives of the Corporation and produced contrary results. Two of the avowed objectives of the Corporation were to make Radio Pakistan more professional and programme-oriented and to provide it freedom in editorial, operational, administrative and financial matters by breaking the fetters of departmental rules, regulations and procedures. The prevalent situation several decades after Corporation is apparently unenviable when compared with pre-Corporation days. The advocates of Corporation probably did not give much thought to the long-term consequences of some of their decisions. These decisions are described briefly as under:

(i) In attempting to free programme developers of routine petty administrative matters, they were stripped of all administrative and financial powers and control. This was done to news staff as well. They were thus made subservient to the administration and finance personnel for their day to day requirements resulting in unnecessary hindrances and frustration for the professionals.

Even a cursory look at the figures will show that the real beneficiaries of Corporation are not the professionals (the term includes programme, news and engineering personnel) but the Administration staff and the Finance and Accounts Wing, the latter being a new creation. In Radio Pakistan there was one Director of Administration of the status of Deputy Controller, 4 Administrative Officers, 10 Superintendents, 59 Assistants, 100 Upper Division Clerks/Head Clerks, 297 Lower Division Clerks, 60 Qasids/Daftaries/Record Sorters and 308 Naib Qasids. This staff was also responsible for finance and accounts and security with very few additional hands, namely 27 Accountants (of the rank of UDC), 1 Security Officer (of the status of an Assistant), 3 Caretakers and 23 Chowkidars. At the end of June 2000, there were 2 Controllers, 4 Deputy Controllers, 9 Administrative Managers, 13 Administrative Officers, 36 Assistant Administrative Officers, 85 Assistants, 172 UDCs, 302 LDCs, 72 Qasids/ Daftaries/Record Sorters and 714 Naib Qasids.

The sharp increase in security staff becomes still more startling when one realizes that the system is being duplicated and police officials are deployed for security at all key points.

As against this, in pre-Corporation days there were on the Programme side two Deputy Directors General, eighteen Directors (including Regional Directors), twenty-nine Assistant Directors including two School Broadcast Officers, forty-seven Programme Organizers, nine Presentation Organizers, ninety-two Producers and thirty-three Presentation Supervisors. They were to man the Headquarters, nine Radio Stations and five Units (External Services, Transcription Service, Staff Training School, Commercial Service and Radio Publications). At the end of June 2000, there was one Director of Programmes, 12 Controllers, 36 Deputy Controllers, 58 Programme Managers, 103 Senior Producers and 152 Producers. Their placements are at the Headquarters, 24 Radio Stations and 7 Units.

The position in the News and Engineering Wings is marginally better. But the fact remains that the creation of new posts and the increase in the number of personnel in the subsidiary wings is much more than in the professional wings, which is not a desirable reflection on the sagacity and foresight of the advocates of the Corporation.

During the second half of 1999 the organizational set-up (scale 5 and above) of all wings of PBC was revised with a view to 'smooth functioning, improving efficiency and removing blockage in promotions resulting in stagnation and frustration'. It was claimed that this had been done in such a manner that the total number of posts in these five scales did not change. In actual fact there is net increase in the number of posts in the Administration Wing (39 per cent) and Engineering Wing (6.5 per cent). In the Finance and News Wings some posts in the lowest cadre were 'surrendered' but posts in higher grades increased correspondingly which means there was an upgrading of all the surrendered posts.

In the Programme Wing, of the 56 'surrendered' posts of Senior Producers, 32 were downgraded to that of Producer while the remaining 24 were upgraded. This is indicative of the ingenuity

Radio Pakistan as Staff Artists. They were the best writers, speakers, artists, actors, musicians and singers, etc. of the time.

iv) Freedom from government rules, regulations and official procedures, that was promised, has not been realized. The Corporation is even more frustrated with red tape than before, particularly in matters of finance.

v) So it is with the promise of freedom in programme policy and editorial judgement. There is a general complaint that Radio is biased in its presentation of all programmes, news, commentaries, discussions, radio reports, etc. It is selective even in the case of entertainment including music, plays, skits and features. Status quo is maintained which puts the institution's credibility at stake.

vi) Professionals and others belonging to auxiliary and supporting services have been treated alike. It can be asked in what way a Naib Qasid, LDC, Assistant, Accounts Clerk, Accountant, Typist, Stenographer, Accounts Officer, or an Administrative Officer in PBC is different from his counterpart in any Ministry or Government department. Posts similar to those in Government departments should enjoy the same pay scales, allowances and other benefits as their equivalents in Government departments. The Corporation should have provided, as was claimed, special pay scales, benefits and facilities for the professionals only as is the case in some autonomous bodies.

In addition to the steps proposed above in this Chapter as well as elsewhere in this book the Corporation should also consider observing the following points in the interest of playing a more beneficial and meaningful role in the new millennium.

a) To follow the primary objective of the creation of Pakistan, i.e. building a society based on *'Amr bil Maroof'* (enforcing virtues) and *'Nahi anil Munkar'* (checking vices). This can best be done not only by broadcasting special religious programmes but by reflecting basic values of Islam in all Radio programmes.

b) To fully avail the autonomy and freedom given to it by the Government and to act as an instrument of the State and not of the Government.

c) To analyse its activities on a regular basis. The guiding principles for planning programmes are: to follow the Constitution, to uphold the democratic values and to speak the truth.

of the Administration, who while appearing to be restructu[red] have in fact substantially increased the number of senior p[osts] whereas the overall position in the Corporation called for a dras[tic] cut across the board in the auxiliary and supporting departmen[ts]. The position of posts before and after restructuring in differe[nt] wings is given below:

Scale	Administration		Finance		News		Engineering		Programme	
	a	b	a	b	a	b	a	b	a	b
5	21	36	71	55	76	65	151	191	120	152
6	13	13	16	26	46	50	342	299	159	103
7	7	9	3	9	28	31	78	112	41	58
8	3	4	4	4	23	26	37	44	30	36
9	2	2	2	2	6	7	10	12	11	12
Total.	46	64	96	96	179	179	618	658	361	361

a = As on 30.06.1999.
b = As on 31.12.1999, or 30.06.2000.

To correct the situation it is necessary that professionals should hold all higher posts at the Headquarters as well as Stations/Units and the number of posts at all levels in the auxiliary and supporting wings should be drastically reduced. Clerical staff may rise up to the level of Administrative Officer provided they are graduates. Otherwise the highest post for them should be that of Assistant. For the ex-cadre and other posts the move-over period should be twenty years to bring it as near the spirit of Government rules on the subject as possible. It created great distortion in PBC. The Government has abolished the move-over system from the current financial year, 2001-02.

ii) Another questionable decision was to abolish the Presentation Section. Even a well-produced programme if not presented properly will fail to impress listeners. Presentation makes all the difference as it can make or mar a programme. A look at the stations in the West shows that the first two positions or priorities are Presentation and Sales, with Production coming third. The Presentation Section has to be revived in the interest of the Corporation as well as the audience.

iii) The provision for staff artists was abolished without making any alternative arrangements. There was a time when practically all 'big names' in the fields of art and literature were associated with

d) To stress the quality of tolerance in people so that the democratic system may flourish.
e) To provide leadership to the masses through uplifting dramatic presentations, music and current affairs programmes.
f) To work towards the eradication of '*jagirdari* system' (landed aristocracy created by the British in undivided India), which has been a major hurdle in national development in any field including radio broadcasting.
g) To take effective steps to increase its regular clientele by improving the quality of its programmes to reflect the preferences and needs of its audiences.
h) To make all efforts to meet squarely the challenge posed by the cultural and political invasion through foreign broadcasting stations.
i) Besides safeguarding the cultural and ideological frontiers, to strive for enrichment and renewal of cultural heritage so that the process of modernization may reconcile the demands of change with those of the continuity of culture. Knowledge of and respect for all cultures be fostered and inter-cultural communication be encouraged.
j) To improve the quality and appeal of its programmes by increasing the signal strength, reviving the presentation section and reintroducing the old practice of operating the panel by the Producer or the Announcer/Compere. This system is followed in most countries of the world and is also more economical.
k) To give special attention to educational programmes. The data systems take CDs into a new realm offering interactive use and a host of educational applications. PBC should make appropriate use of this new technology.

In short, Radio Pakistan is to prepare society for the new millennium by promoting Islamic values, emphasizing the ideology of Pakistan, strengthening national cohesion, avoiding all sorts of '*munkarat*' (vices and forbidden things) and meeting the fierce challenges of the modern world.

Let us end this book with a supplication to Almighty Allah that in the discharge of its functions and obligations and its communication with the people, Radio Pakistan may always remain wedded to its motto.

"قُولُوا لِلنَّاسِ حُسْناً"

21

STATIONS

1. ISLAMABAD STATION

Inauguration	07.05.1977
Transmitter Power	1,000 KW MW and 2 KW FM
Radius	400 kilometres
Frequency	585 kHz (MW)
	101 mHz (FM)
Languages	Urdu and English.
Transmission, daily average	39 hours and 30 minutes.

Before shifting to the National Broadcasting House on 17 May 1984 the Islamabad Station was almost entirely dependent on Rawalpindi Station. Producers and studios were borrowed from that Station to make a start. In the beginning the transmission timings were from 6.45 a.m. to 11 a.m. Later two hours were added from 7 p.m. to 9 p.m. Now the Station goes on air for about 39 hours 30 minutes daily including 22 hours of FM Service.

The Islamabad Station is somewhat different from other Stations of PBC. It is a national channel and the tone and contents of its programmes have a national approach. It has no orchestra and does not produce any of its own music but the best music produced elsewhere in the country is broadcast from this Station. The religious programme 'Hayya Alal Falah' and DJ 'Subh-i-Pakistan' are broadcast on the national hook-up. The five-hour *Sautul Quran* channel is broadcast from Islamabad since 26 January 1998.

Radio Forum in which public representatives, Ministers, Senators, Chairmen of Parliamentary Committees, heads of Ministries/Divisions/Departments/autonomous bodies, etc. face a panel of interviewers, has been on the national network every week since October, 1993. All-Pakistan *Naat* competition, *Husn-i-Qirat*, Seerat Conference and election coverage are 'speciality' programmes for this Station. It originates a programme in English 'Islamabad Calling' for about an hour every evening since 1986 for foreign missions and English speaking communities to inform them about Pakistani culture, folk and classical music. Being a central station, located in the capital of the country, it has more OB activities than any other station and there is a regular section for VIP coverage.

Islamabad is one of the three stations that have an FM Channel for music, light programmes and city service. It was started on 8 April 1993 for two hours, which gradually increased to eight hours. It was revamped on 1 October 1998 with drastic changes in planning, presentation and style. Due to its popularity the duration of FM 101 was raised to 17 hrs a day from 1 January 1999 and further to 22 hours from 10 February 2000. It is very popular with listeners.

2. Lahore Station

Inauguration	16.12.1937
Transmitter Power	5 KW MW (Commissioned on 16.12.1937)
	100 KW MW (Commissioned on 16.10.1965)
	50 KW MW (Second channel)
	2 KW FM (started on 08.4.1993)
Radius and Frequency	Channel – I = 240 KM, 630 kHz.
	Channel – II = 120 KM, 1080 kHz.
	FM = 30 KM, 101 mHz.
Transmission, daily	46 hours. average
Languages	Urdu and Punjabi.

In a sense radio broadcasting started in Lahore in 1928 when the YMCA set up a 100 watt medium wave transmitter which continued up to 1 September 1937. Lahore Station of All India Radio formally started service on 16 December 1937. It came into the ambit of Pakistan Broadcasting Service on 14 August 1947. The Station shifted to its present building in 1965.

In the tragic aftermath of Independence, Lahore Station did an excellent job of uniting divided families and missing persons by airing about 37,000 SOS messages from 27 August 1947 to end of March 1948. At the same time it started programmes to bring solace to displaced persons and for morale boosting and nation-building. Similarly it played a very effective role during the 1965 War and produced many programmes and songs during that crisis.

Starting from one transmission of 4-5 hours, the Station had three transmissions at the time of Independence of about nine hours. Now it is on air for about 46 hours a day on two medium wave channels and one FM Service of 22 hours. All categories of programme are broadcast for general consumption as well as for special audiences like women, children, students, youth, troops, the sick and the handicapped, businessmen, farmers and the rural population. During emergencies and natural calamities the Station makes special efforts to cover rescue and relief operations. Lahore Station has had a galaxy of writers, scholars, academicians, theologians, playwrights, musicians, drama and music artists, comperes and announcers associated from the beginning. After Independence, Ahmad Salman became the first Station Director.

3. MULTAN STATION

Inauguration	21.11.1970
Transmitter Power	120 KW MW
Radius and Frequency	250 KM and 1035 kHz.
Transmission, daily average	15 hours 45 minutes
Languages	Urdu and Saraiki

Multan, also known as Madinatul Aulia and famous for its *Kafis*, came on the air on 21 November 1970 (20 Ramazan 1390 AH) with a transmission of seven and a half hours. It was inaugurated by the then Minister for Information and Broadcasting, Nawabzada Sher Ali Khan. Different types of programmes to cater to different segments of society were started from the beginning. It is the first station to broadcast programmes in Saraiki.

Commissioning of Multan Station provided an opportunity to the local talent to prove their worth and a number of artists attained nationwide recognition. Three of them, Iqbal Bano, Pathane Khan and Surayya Multanekar, are recipients of the President's Award for Pride of Performance. In 1974, Multan started producing regional music in addition to national music. Its programme 'Music in Multan' won international recognition in 1993 at the fourth Radio Shanghai Music Festival. During the devastating floods, it broadcast continuous transmission for 72 hours in view of the gravity of the situation. It was adjudged to be the best station of Radio Pakistan for 1987.

Saraiki was limited to *kafi, masnavi* and *dohre*. Radio helped introduce short story, essay, review and critique, so that literature in prose was also added in Saraiki. A play in Saraiki was put on the air from Multan for the first time in Pakistan. Shamsuddin Butt was the first Station Director.

4. Faisalabad Station

Inauguration	15.09.1982
Transmitter Power	250 watts
	10 KW MW (since 01.10.1989)
Radius and Frequency	60 KM and 1476 kHz.
Transmission, daily average	12 hours 5 minutes.
Languages	Urdu and Punjabi.

Faisalabad is the third biggest city of Pakistan. The Station was inaugurated by the Minister for Information and Broadcasting, Raja Zafarul Haq, on 15 September 1982 (26 *Ziqaad*, 1402 AH). It started as a model city station with emphasis on public service programmes and was on the air in the evening from 3.00 p.m. to 9 p.m. Its transmission was shifted to the morning in December, 1982 and was broadcast from 5.45 a.m. to 1.20 p.m. for over six years. In July 1989, transmission time was changed to evening from 4 p.m. to 11.10 p.m. On public demand it was again shifted to morning from 5.45 a.m. to 1.20 p.m. in October, 1989. On 1 February 1995 the transmission was extended up to 5.15 p.m. The extended time from 2 p.m. to 5 p.m. was filled by relaying Islamabad programmes. Towards the end of the year the transmission was divided into two, morning (0545-1204) and evening (1330-1715). Now its programmes can be heard from 7.55 a.m. to 8 p.m.

During the September 1992 floods, this Station played a vital and effective role. The affected communities were repeatedly warned of the impending danger and advised to move to safer places. Teams were sent to Trimmu Headworks and other strategic places and people were kept constantly informed of the latest flood situation. These included reports sent by PBC representatives from various locations in Jhang District. Constant liaison was maintained with the local authorities and the General News Room at Islamabad was kept updated. The OB teams reached various places even at the risk of their lives and sent reports that were instantly broadcast from Faisalabad.

To improve the quality of relay programmes satellite receiving equipment was installed at Faisalabad Station on 4 October 1994 and from the very next day it started relaying national hook-up programmes through this system so that the relayed programmes could be heard without any interference and noise. Muhammad Akram Chaudhary was the first Station Director.

5. BAHAWALPUR STATION

Inauguration	18.08.1975
Transmitter Power	10 KW MW
Radius and Frequency	80 KM and 1341 kHz.
Transmission, daily average	11 hours and 25 minutes
Languages	Urdu and Saraiki.

Bahawalpur Station was formally inaugurated on 18 August 1975 though test transmission had started on 1 July 1975. It is the first station where the transmitter was installed within the premises of the Broadcasting House. In the beginning it had two studios. Later, an office block and one studio were added in 1988 that was formally inaugurated on 11 July 1990. In the beginning its transmission was of seven hours from 1.50 p.m. to 9 p.m. Later it was extended to 11.10 p.m. so that the total transmission was of 9 hours 20 minutes, on weekdays and 8 hours 20 minutes on Fridays. In 1999 morning transmission was added from 8.50 a.m. to 11.05 a.m. (i.e. for about 2 hours 15 minutes).

Its social service and city service are very effective. During the floods, Bahawalpur Station has rendered exemplary service. Following the devastating floods of 1988 it had to run the transmission for 36 hours. In view of the danger to Panjnad Headworks during the 1992 floods Bahawalpur Station broadcast direct from the Headworks for several hours. The rural and farm forum programmes give useful information about crop, sheep, goat and other livestock diseases. Some of its programmes have won laurels from ABU, UNICEF, SAARC, and PBC Headquarters. A number of writers and performers associated with this Station have been recognized on the national and even on international level. The Station also experimented with *Kafis* in chorus, thematic plays and features and running commentary on local sports events. Ahmad Sajjad Tirmizi was the first Station Director.

6. Rawalpindi Station

Inauguration	01.09.1950
Transmitter Power	100 watt MW
	10 KW MW (Since 21.5.1953 replacing the low-power transmitter)
Radius	80 kilometres
Frequency	1152 kHz
Transmission, daily Average	15 hours and 45 minutes
Languages	Urdu, Pothohari and Kashmiri.

Rawalpindi Station has a unique distinction in that its inaugural announcement was made by Z.A. Bokhari himself: '*Yeh Radio Pakistan Rawalpindi Hai. Assalamo-Alaikum*'. Another point of interest is the number of the building housing the Broadcasting House. One of the reasons for setting up Rawalpindi Station was to counter the baseless and malicious Indian propaganda against Pakistan about Kashmir. Because of the association of the number with the rifle it was surmised that it would prove a befitting answer to Indian propaganda. The Station was inaugurated by the Minister for Kashmir Affairs, Mushtaq Ahmed Gurmani.

After the war of September 1965, Rawalpindi Station broadcast interviews of about 1,100 Indian prisoners of war. Weather bulletins for mountaineers were also broadcast for the benefit of expeditions. It is probably the first station to present music programmes on the national hook-up. This Station broadcast special services in Kashmiri, Balti and Shina for listeners in northern areas. The Services in Shina and Balti along with staff were transferred to Gilgit and Skardu Stations when they came on the air on 2 and 16 April 1979 respectively. The service in Kashmiri is continuing on short wave. '*Chalta Phirta Microphone*' was for the first time introduced in November 1977. Mahmood Nizami was the first Station Director.

7. Karachi Station

Inauguration	14.08.1948 (Intelligence School Barrack).
	16.07.1951 (Present BH).
Transmitter Power	100 watt (from 14.08.1948 to 31.10.1948)
	10 KW MW (from 01.11.1948) Channel- II.
	100 KW MW(from 23.3.1977) Channel – I
	5 KW FM.
Radius	100 kilometres.
Frequency	Channel – I = 828 kHz
	Channel – II = 639 kHz
	FM = 101 mHz.
Transmission, daily average	44 hours and 30 minutes
Languages	Urdu, English, Sindhi and Gujarati.

As there was no radio station in the capital city of the new nation an emergency station with radiating power as low as that of an electric bulb used in a house, was inaugurated on 14 August 1948 with skeleton staff and basic equipment and furniture. The makeshift transmitter was replaced with a 10 KW MW transmitter on 1 November, i.e. about two and a half months after the inauguration. Two short wave transmitters of 50 KW each were commissioned towards the end of 1949, which enabled starting of External Services and provided a link between East and West Pakistan. Programmes in Sindhi, Balochi and Gujarati from Karachi Station were started on 14 August 1948, 25 December 1949 and 10 March 1952 respectively. A programme in Kashmiri was also broadcast from Karachi. Western music began from 1 May 1949 for one hour daily.

The Station shifted from the Intelligence School Barracks to the present site on 16 July 1951. A new multi-storey Broadcasting House is almost complete in the Civic Centre. It will accommodate almost all the offices of Radio Pakistan functioning in Karachi.

This Station introduced Drama Festival, Students' Week and *Alami* Sports Round-up (1985). The National Anthem was rehearsed and recorded in the studios of Karachi Station. It serialised novels for the first time. Qamar Ali Abbasi, a well-known dramatist, adapted for Radio and produced such novels as *Anwar, Afshan* and *Nilofer*.

Karachi Station has been fortunate in the choice of personnel. It had a galaxy of scholars, intellectuals, writers, poets, broadcasters, artists, etc. Sajjad Sarwar Niazi was the first Station Director.

8. Hyderabad Station

Inauguration	17.08.1955
	28.12.1962 (present BH)
Transmitter Power	1 KW MW (from 17.08.1955 to 24.12.1959)
	10 KW MW (from 25.12.1959)
	120 KW MW (from 04.02.1971)
Radius and Frequency	Channel-I = 120 KM and 1008kHz.
	Channel-II = 40 KM and 1098kHz.
Transmission, daily average	21 hours 45 minutes.
Languages	Urdu and Sindhi.

Hyderabad Station began with a 1 KW medium wave transmitter on 17 August 1955. This transmitter was replaced with a 10 KW medium wave transmitter on 25 December 1959. A 120 KW MW transmitter was commissioned on 4 February 1971 as Channel-I. The 10 KW transmitter is being used as Channel-II. It shifted to the present Broadcasting House in December 1962, which was inaugurated by President Ayub Khan on 28 December 1962.

The Station arranged the Mirza Qaleej Beg, Hasanali Effandi, Allama Umar ben Muhammad Daudpota and Allama I.I. Qazi Memorial Lectures. Hyderabad was the first Station to celebrate its inauguration day anniversary. It successfully experimented teaching painting through the medium of sound. It started the programme 'Para ba Para' during *Ramazan* which became a national hook-up programme the next year. It also organized one-week specials every month continuously for several years. These included weeks for women, children, youth, students, rural folks, literature, regional sports, farming, local seasonal festivals, etc.

During the 1965 War some of the programmes of Hyderabad Station became very popular. An outstanding achievement of Hyderabad Station relates to transforming a very backward village Phalkara into a model village within a period of two years with the assistance and cooperation of local administration and various Provincial departments. Zafar Husain was the first Station Director.

9. Khairpur Station

Inauguration	03.03.1981 (Experimental)
	21.03.1983
Transmitter Power	100 KW medium wave
Radius and Frequency	110 KM and 297 kHz.
Transmission, daily average	15 hours 45 minutes.
Languages	Urdu and Sindhi.

Khairpur Station with 100 KW MW transmitter started two hour programmes in a temporary building on 3 March 1981. A little later when the Broadcasting House was established on 19 November 1981, it started relaying programmes of Hyderabad Station. The Station was formally inaugurated on 21 March 1983 and it started radiating originated programmes for 8 hours and 15 minutes. This continued up to the inauguration of the existing Broadcasting House by the then Prime Minister, Muhammad Khan Junejo. From the same date it began its regular transmission of over 15 hours.

A result-oriented programme is 'Jot Jalandi Rahey' which aims at providing needy students complete set of textbooks free of cost. Several thousand students have benefited through this programme. It provided impetus to social, cultural, literary and educational activities in the area as well as creating and discovering new talent in different fields including music.

SOS messages and appeals for blood donations are broadcast whenever needed. A programme about handicapped persons is a regular feature every week. A number of students have written a thesis on programmes of Khairpur for their Masters degree in journalism/ mass communication, Urdu and Sindhi literature. Munir Ahmad Soomro was the first Station Director.

10. LARKANA STATION

Inauguration	24.09.1995
Transmitter Power	1 KW FM Stereo.
Radius and Frequency	35 KM and 101 mHz
Transmission, daily	6 hours 22 minutes.
Languages	Urdu and Sindhi.

Larkana, the fourth Station in Sindh, was inaugurated on 24 September 1995 and on the same date the foundation-stone was laid for the Broadcasting House which will have a 100 KW medium wave transmitter. The transmission as usual begins with religious items. The national hook-up chunk of religious programmes from 1.04 p.m. to 1.30 p.m. is relayed from Islamabad. The weekly programme 'Tarashey' is very popular. It is based on contributions received from listeners. Khwaja Imdad Ali was the first Station Director.

11. PESHAWAR STATION

Inauguration	16.07.1942
	28.04.1985 (Shifting to new BH)
Transmitter Power	300 KW MW (since 28.04.1985)
	100 KW MW (since 13.11.1993) replacing 10 KW MW transmitter of 1942)
	10 KW SW (since 16.01.1960)
Radius and Frequency	Channel-I = 250 KM and 540 kHz.
	Channel-II = 150 KM and 1377 kHz.
Transmission, daily average	26 hours
Languages	Urdu, Pushto, Chitrali, Hindko, Kohistani, Gojri.

The formal date of inauguration of the Station is 16 July 1942, but a beginning was made in 1935 when in January of that year the Marconi Company offered the NWFP Government a 250 watt transmitter on loan together with some sets for community listening particularly in Hujras. On 1 April 1936 Peshawar became a regular broadcasting station. With the outbreak of the Second World War, the Station was downgraded to the status of a relay station. However, after three years in 1942 a formal Broadcasting House with four studios was constructed along with erection of a 10 KW medium wave transmitter replacing the old 250 watt transmitter. After Independence a 10 KW short wave transmitter was added in 1960, a powerful 300 KW medium wave transmitter on 28 April 1985 and a 100 KW medium wave transmitter in November 1993. The present BH was inaugurated by President Ziaul Haq on 28 April 1985 (7 *Shaban* 1405 AH)

Some special programmes of Peshawar Station include: a Pushto External Service programme daily for Tribals and Afghanis educative plays in Pushto and Persian in connection with reconstruction in Afghanistan in collaboration with some international agencies; these plays are rebroadcast from Quetta Station and BBC, PBC school broadcasts for primary level students, for the present teaching English language. After Independence S.S. Niazi became the first Station Director.

12. Dera Ismail Khan Station

Inauguration	15.01.1981
Transmitter Power	10 KW MW
Radius and Frequency	80 KM and 1404 kHz
Transmission, daily	7 hours 15 minutes.
Languages	Urdu, Saraiki and Pushto.

Dera Ismail Khan started with a transmission of 6 hours 15 minutes. Three hours and ten minutes programmes were originated and for the remaining three hours five minutes it relayed programmes of Peshawar and Islamabad Stations. The transmission has been extended by about an hour and almost all programmes originate at the Station except for one hour from 8 p.m. to 9 p.m. All types of programmes are broadcast. A new drama is put on air once a month. Skits and short plays are broadcast in Saraiki and Pushto. Programmes for farmers in Saraiki were broadcast from the beginning while the Pushto programme was introduced in 1986. Literary and cultural activities of Bhakkar District are also covered by D.I. Khan Station in a weekly report. Commercial programmes including city service were started in 1987. Programmes for women and children are presented in all the three languages of the Station. Yaqoob Bangash was the first Station Director.

13. ABBOTTABAD STATION

Inauguration	21.03.1991
Transmitter Power	250 watts medium wave.
Radius and Frequency	25 KM and 1602 kHz.
Transmission, daily	5 hours 30 minutes.
Languages	Urdu and Hindko.

In the beginning, Abbottabad Station originated programmes for one hour and 15 minutes. In October 1993 this was raised to two hours and 15 minutes. For the remaining period of its transmission of five and a half hours it relays programmes from Islamabad and Peshawar Stations. Its transmission begins at 1.45 p.m. with a 15 minute religious programme. From 2 p.m. to 4 p.m. it relays Islamabad and for the next one hour the relay is from Peshawar. At 5.15 p.m. begins its Hindko programme and at 6.05 p.m. an Urdu programme. Both the composite programmes are popular among the listeners.

The Station broadcasts programmes of general nature and for such audiences as children, women, students, farmers and sportsmen. Social and cultural activities in Hazara are covered in a special programme. It radiates special programmes on national days, anniversaries and festivals. During the 1992 floods this Station broadcast daily reports on the status and relief operations. An important contribution of Abbottabad Station is the patronage given to Hindko language. Without this patronage the Hindko heritage would probably have never come to light. Nisar Muhammad Khan, Programme Manager, was the first Station Director.

14. CHITRAL STATION

Inauguration	01.08.1993
Transmitter Power	250 watts medium wave.
Radius and Frequency	15 KM and 1584 kHz.
Transmission, daily	4 hours 27 minutes.
Languages	Urdu and Chitrali (Khwar)

The foundation-stone of Chitral Station was laid on 15 October 1991 and it began its transmission on 1 August 1993 at 4.50 p.m. At that time it originated one hour's programme in Urdu and Chitrali and relayed one hour's programme from Peshawar. The transmission time was increased by one hour. It now originates Chitrali programmes for one hour 15 minutes and Urdu programmes for over two hours. It relays Chitrali programmes including a Chitrali news bulletin from Peshawar for 45 minutes. (It may be mentioned here that Peshawar Station started a 5-minute weekly programme in Khwar on 7 November 1965, which is now broadcast daily for 45 minutes).

Among the important events covered by the Station are the 5-day International Hindu Kush Cultural Conference in August 1995, Shandur Polo Tournament and traditional festivals of the Kalash Valley. Besides providing exposure to local intellectuals, scholars and artists the Station made special efforts for regional cultural, literary and music development of the area. Faqir Husain Sahir, Programme Manager, was the first Station Director.

15. QUETTA STATION

Inauguration	17.10.1956
Transmitter Power	1 KW MW (up to 31.05.1961)
	10 KW MW (from 01.06.1961)
	10 KW SW (from 17.10.1962)
	150 KW MW (from 25.04.1974)
Radius	35 KM and 100 KM.
Frequency	756 kHz – Channel – I
	855 kHz – Channel – II.
Transmission, daily average	32 hours 15 minutes.
Languages	Urdu, Balochi, Pushto, Brahvi and Hazargi.

Quetta Station, housed in a private building, started with six hours transmission in Urdu and Balochi on a 1 KW medium wave transmitter. Its power was raised to 10 KW and then to 150 KW. A 10 KW short wave transmitter was also added. With the commissioning of 150 KW transmitter the transmission duration was raised from 17 hours to about 30 hours. The present premises were occupied in 1962. Four fully equipped studios and an office block were added in 1980.

Quetta Station played an effective role in the development of Balochi, Pushto and Brahvi languages. Among the outstanding activities of the Station are collection and preservation of Balochi and Brahvi folk music and interviews of those who took part in the Pakistan Movement, poetic translation of Allama Iqbal's works in Balochi and Brahvi and of Jam Durak's poetry into Urdu poetry. Another contribution is in the field of education of women. Because of local

17. Turbat Station

Inauguration	04.01.1981
Transmitter Power	250 watts.
Radius and Frequency	25 KM and 1584 kHz
Transmission, daily	5 hours 30 minutes.
Languages	Urdu and Balochi.
	1800 – 1900 = Balochi.
	1900 – 2000 = Urdu
	2000 – 2100 = Relay Islamabad.
	2100 – 2300 = Balochi.

At the time of its establishment Turbat was the second Station of PBC in Balochistan. Its importance lay in the fact that practically no station of Radio Pakistan was able to reach the area. It was set up in pursuance of the policy of the Government to reach the remote areas and provide them a sense of involvement in the affairs of the nation. The Station broadcasts different types of programme to cater to the needs of various segments of society. Its programmes meet the needs of the local population in the fields of culture, education, agriculture, literature and entertainment. Maroof Ali was the first Station Director.

18. Sibi Station

Inauguration	01.10.1989
Transmitter Power	250 watts.
Radius and Frequency	10 KM and 1584 kHz
Transmission, daily	3 hours 13 minutes.
Languages	Relays Islamabad.

Sibi Station was inaugurated with a low power village broadcaster transmitter of 250 watts on 1 October 1989 as a relay station. Except for the opening and closing items of about 13 minutes the Station relays programmes of Islamabad Station from 1 p.m. to 4 p.m.

19. Loralai Station

Inauguration	20.07.1996
Transmitter Power	10 KW medium wave.
Radius and Frequency	35 KM and 1251 kHz

tribal customs literacy among women is very low. Radio because of its reach and advantage of being listened to while working, has been very helpful and effective in the uplift of women, providing not only useful information about household affairs but also helped in raising the national literacy level. Quetta was adjudged the best Station for the year 2000. K.G. Ali was the first Station Director.

16. Khuzdar Station

Inauguration	17.06.1981
Transmitter Power	250 watts.
	300 KW MW (since 03.09.1991)
Radius and Frequency	300 KM and 567 kHz
Transmission, daily	6 hours 18 minutes.
Languages	Urdu, Brahvi and Balochi.

The Station was inaugurated on 17 June 1981 with a 250 watt transmitter with coverage within a radius of about 20 kilometres. On the installation of the 300 KW medium wave transmitter (fabricated and erected by PBC engineers) on 3 September 1991 and increase in its coverage area, an advisory committee was constituted to make recommendations about programmes to be broadcast. The greatest stress was laid on increasing the rate of literacy especially among women. Special programmes were broadcast urging parents to send their children including girls to schools. A number of schools both in public and private sectors started functioning. Philanthropists and the rich were urged to help poor and needy students in getting education at different levels.

The second major recommendation related to health and hygiene. Talks, discussions and interviews were scheduled for the purpose with tangible results. 'Chalta Phirta Microphone' in Urdu and 'Sailani' in Brahvi were started to highlight day to day problems including health and hygiene. These programmes soon became popular and served their purpose. Similarly development of agriculture and livestock breeding also received special treatment that helped in increasing yield per acre. Local talent in various fields was discovered and demands made by different segments of society to introduce special programmes were duly honoured. A. Sami Qureshi was the first Station Director.

Transmission, daily	4 hours 30 minutes.
Languages	Urdu and Pushto.

Loralai, the fifth radio station in Balochistan, was inaugurated on 20 July 1996. At that time it relayed programmes of Quetta Station from 5 p.m. to 7 p.m. and of Islamabad Station from 7 p.m. to 9 p.m. Now only two hour programmes are relayed: 1800-1900 hours from Quetta and 2000-2100 hours from Islamabad. Two hours and 30 minutes programmes are originated by Loralai itself. Agha Muhammad Kasi was the first Station Director.

20. ZHOB STATION

Inauguration	21.04.1998.
Transmitter Power	10 KW medium wave.
Radius and Frequency	60 KM and 1449 kHz
Transmission, daily	4 hours 20 minutes.
Languages	Urdu and Pushto.

Zhob, the sixth Radio Station in Balochistan and 24th in the country, was inaugurated on 21 April 1998. PBC engineers fabricated the audio equipment installed at the Station and the transmitter. The Station broadcasts all types of programmes e.g. religious, cultural, educational, agricultural, other information programmes, drama, feature, music, etc. catering to the needs of all segments of society. Talent hunting, at this Station, like all other stations, is a regular feature. Agha Muhammad Kasi was the first Station Director.

21. GILGIT STATION

Inauguration	02.04.1979
Transmitter Power	250 watts (up to 30.09.1989)
	10 KW MW (since 01.10.1989)
Radius and Frequency	50 KM and 1512 kHz
Transmission, daily	7 hours
Languages	Urdu, Shina, Brushaski and Wakhi.

The Station started in a rented house with a 250 watt transmitter which could be heard up to about 10 miles. Its broadcast hours were from 3 p.m. to 9 p.m. presenting different types of programmes in Urdu and Shina. Programmes in Brushaski were introduced from 8

October 1984 and in Wakhi from 1 November 1996. With the installation of 10 KW MW transmitter in 1989 it now reaches almost the whole of the northern areas including Azad Kashmir. During the night it can be heard in NWFP and many parts of the Punjab.

Gilgit Station has played a very constructive role in projecting the social and cultural life in the inaccessible valleys of Gilgit region, recording and preserving folk music of the area, popularizing polo, skiing and mountaineering by broadcasting special weather bulletins and introductory programmes, creating a sense of patriotism and involvement among the people and helping farmers in collaboration with the local agriculture departments, taking timely steps during natural calamities like earthquakes and torrential rains. An important contribution of Gilgit Station is in raising the level of Brushaski from a dialect to a language. There was no distinction between *Hamd, naat* and mystic poetry in the region. Gilgit Station took special interest in the matter and recorded a large number of compositions on the three subjects. Salim Iqbal was the first Station Director.

22. SKARDU STATION

Inauguration	16.04.1979
Transmitter Power	250 watts (up to 22.11.1982)
	10 KW MW (from 23.11.1982)
Radius and Frequency	25 KM and 1557 kHz
Transmission, daily	7 hours
Languages	Urdu and Balti.

Skardu Station is the highest Radio Station in Pakistan, being located about 7,500 feet above sea level. It started broadcasting in Urdu and Balti. A Balti programme of 30 minutes was first broadcast from Rawalpindi-III Station in 1949. Later it was shifted to Rawalpindi Station from where it continued to be originated till the inauguration of Skardu Station. One of its important contributions has been turning the Balti dialect into a language. Balti lacked good *ghazals*, had no concept of national/patriotic songs and musical features. These were introduced for the first time from Radio Pakistan. Another important contribution was the broadcast of a series of 50 talks on the social, cultural and literary history of Baltistan and interviews of freedom fighters who against heavy odds successfully fought the Dogra Army. With the installation of 10 KW medium wave transmitter, Skardu is able to reach the whole of the Northern Areas as well as parts of Kargil, Laddakh and Kashmir. The Station rendered exemplary service during avalanches, lightening, landslides and floods. Haseeb Ahmad Khan was the first Station Director.

23. RAWALPINDI-II STATION

Inauguration	15.10.1960
Transmitter Power	1 KW SW
	10 KW MW (from 29.04.1976)
	150 KW MW (from 2.10.1983)
Radius and Frequency	100 KM and 792 kHz
Transmission, daily average	12 hours 20 minutes.
Languages	Urdu, Pahari, Gojri, Kashmiri and English.

The Station began its transmission on 15 October 1960 on a one kilowatt short wave transmitter with the primary objective of propagating the cause of Kashmiris suffering under the Indian occupation forces in addition to keeping the people informed of the Government's activities for their prosperity and development as well as reminding them of their ideological direction. To reinforce its voice and coverage area a 10 KW and another 150 KW medium wave transmitters were installed so as to expose the Indian despotic rule and tyranny against the peace-loving Kashmiris of Jammu and Kashmir who have been struggling for freedom from Indian occupation.

The Station broadcasts all types of programme but the liberation movement in the Indian-occupied State of Jammu and Kashmir gets first priority. However, in deference to the subjugated position of the people of occupied Kashmir no music or entertainment programme is aired by this Station. Devotional and national songs are broadcast on regular basis. There is a weekly programme 'Voice of Kashmir' of 35 minutes in English also about India's sinister designs in the occupied State and at the same time highlighting the right of self-determination of the Kashmiris. Masud Qureshi was the first Station Director.

24. Rawalpindi –III Station

Inauguration	16.04.1948
Transmitter Power	100 KW short wave.
Transmission, daily average	12 hours 03 minutes.
Languages	Urdu, Kashmiri, Gojri and Pahari.

The Station started broadcasting on 16 April 1948 with a low power (500 watt) short wave transmitter. Its power was raised in 1949 and again in 1968 to 100 KW SW. Its main purpose is to uphold and propagate the cause of Kashmir and expose the Brahminic fanaticism and tyranny of Indian occupation forces against the Kashmiris, keep the people informed of Indian designs and conspiracies, highlight the efforts being made by the people and heroic deeds of the freedom fighters, to counter the venomous and baseless propaganda by Indians and to give a true picture of happenings in occupied Kashmir. Mahmood Nizami was the first Station Director.

Appendix–I

Press Communiqué of Government of British India

The Hon'ble Sir Syed Sultan Ahmad, Member Incharge of the Department of Information and Broadcasting, met in special conference the representatives of the Anjuman-i-Taraqqi-i-Urdu consisting of Dr. Abdul Haq, Dr. Abdul Sattar Siddiqui and Pandit Brij Mohan Dattatriya Kaifi and of the All India Hindi Sahitya Sammelan, consisting of Sri Pandit Sampurnanand, Sri Mauli Chandra Sharma and Swami Bhadant Anand Kausalyan and the Hon'ble Syed Husain Imam, Member Council of State, to arrive at a satisfactory solution of the Hindi-Urdu language problem in AIR.

After giving very careful consideration to the views expressed by the representatives of the two bodies and Mr. Syed Husain Imam, the Government of India have taken the following decisions:

I. News, News Commentaries and Announcements

(1) There should not be separate Urdu and Hindi broadcasts of news, news commentaries and announcements, the language used being of the simplest variety, which is understood by the vast majority of the listeners in Hindustani speaking areas.

(2) Where there is no appropriate Hindustāni word, and a foreign word has to be imported, the basis of selection from an indigenous language, Persian, Arabic or English should be the widest intelligibility of that word in Hindustani speaking areas.

(3) Whenever foreign words are used their deflections should conform to the grammar, not of the language from which they are taken but of the grammar of Hindustani (e.g. if the word "station" were taken from English and adopted by Hindustani, the plural form should be "stationon" and not "stations"; similarly "qaida" and "qaide" and not "qaida" and "qawaid"; "fikr" and not "afkar", etc.) No word should be deliberately rejected merely on the ground that there is another word which expresses the same meaning. This should, however, be subject to the general principle that the more widely understood words are to be preferred.

(4) As regards pronunciation,

a) the most current form of pronunciation should be preferred to the purist literary and etymological form,
b) the name of person, as far as possible, be pronounced as he pronounces it himself, and
c) for names of places, rivers, etc. the most widely current form should be preferred.

II. Composite Programmes, e.g. Women's and Children's Programmes

The atmosphere of these programmes should not be exclusively allied to any one particular community.

III. Talks

Talkers should generally be given latitude to use the style of language to which they are accustomed, but attempts should nevertheless be made to persuade them to use as simple a language as possible so that the talk reaches the largest number of listeners.

In the selection of talkers every effort should be made to ensure equitable representation of talent drawn from all sections and communities and to reflect cultural and social needs and interests of the areas served by each station.

Information and Broadcasting Department
New Delhi, February 14, 1945.

....

The above press communique was laid on the table of the Indian Legislative Assembly on February 14, 1945.

....

Appendix-II

Press Note of Government of British India

At the invitation of the Hon'ble Sir Akbar Hydari, Member for Information and Broadcasting, Committee consisting of the Hon'ble Rai Bahadur Sri Narain Mahtha, Member, Council of State, Nawab Siddique Ali Khan, M.L.A. (Central), Dr. Zakir Hussain and Dr. Tarachand, met on the 26th of January 1946 to advise the Government of India on the following questions:

(i) Should the Hindustani news of All India Radio continue to be broadcast in a common language or should it be broadcast separately in Hindi and Urdu.

(ii) If in a common language, what steps should be taken for the selection of vocabulary which should be satisfactory and generally acceptable?

(iii) What programme composition would achieve a fair representation of Hindi and of Urdu in spoken-word items other than news?

All the previous discussions on the Hindi-Urdu question which had taken place under official auspices, including the debates in the Legislature, were placed before the Committee and the difficulties that had arisen on account of the rival demands of the various sections of the people regarding the language of the Hindustani news bulletins and the composition of the Hindustani programmes of All India Radio were brought to their notice.

2. So far as the language of the Hindustani news bulletins is concerned, it was pointed out to the Committee that during the last ten years or so the linguistic style of the Hindustani news bulletins had been fluctuating either from purely practical considerations or in response to the varying demands of the public. The language of these bulletins was, therefore, to be regarded as an experiment which had been in progress for some years, and the question for consideration now was whether this experiment should be continued and, if so, on what lines. In this connection, All India Radio had compiled a Lexicon of about 8,000 English words commonly used in news, together with their Hindi and Urdu equivalents and suggestions for simple Hindustani synonyms, with the intention of circulating it to experts in the country for their comments on the simple synonyms suggested. The Lexicon was placed on the table.

3. After a full discussion, the Committee unanimously came to the conclusion that the use of Hindustani as the common language for news bulletins though not free from difficulties should not be given up without a further attempt at arriving at a generally acceptable vocabulary. For this

purpose the Committee recommended that a Standing Advisory Committee should be set up to advise the Director General, All India Radio, on the choice of vocabulary for Hindustani and that the All India Radio Lexicon, which was a very valuable and constructive compilation, should be circulated to suitable experts in the country for comments which, when received, should also be considered by the Standing Advisory Committee suggested above.

4. One member, while agreeing with these recommendations, proposed that of the four Hindustani bulletins broadcast daily, two should be given in Hindustani, one in Urdu and one in Hindi. This was not agreed to by the rest of the Committee as they feared that the existence of an Urdu and a Hindi bulletin side by side with bulletins in a common language would tend to defeat the purpose of the latter reducing its chances of success as an experiment and the Urdu and Hindi bulletins, because of their separatist character, would tend to become literary and would, therefore, be less widely intelligible.

5. For spoken-word programmes other than news, such as talks, plays, poetry recitations, etc. the Committee unanimously recommended that items in Hindi and items in Urdu should continue to be broadcast in spoken-word programmes other than news as it was most desirable to help and not to hinder the literary development of the two languages and that the question of fair and adequate representation, as between the two, should be referred to the Standing Advisory Committee mentioned above, as the matter would require detailed examination.

6. As for the composition of the Standing Advisory body, the Committee recommended that it should consist of experts representating the Anjuman-e-Taraqqi-e-Urdu, the All India Hindi Sahitya Sammelan and the Hindustani Prachar Sabha, with temporary members to be co-opted by Director General, All India Radio as and when the questions coming up before the Committee required it.

7. It was also suggested that as an auxiliary to the Lucknow Station of All India Radio, a studio should be installed at Allahabad where spoken-word items should be arranged for inclusion in the programmes of the Lucknow Station as it was felt that, pending the installation of more radio stations, this arrangement would enable the Lucknow Station to utilize the talent in the Eastern part of the United Provinces, more adequately than it could at present, in view of the long distance involved.

8. The Government of India have decided to accept all the recommendation of the Committee. Steps are being taken to set up a Standing Advisory Committee on the lines suggested by the Committee, to circulate the All India Radio Lexicon to suitable experts in the country for eliciting opinion and to install an auxiliary studio in Allahabad as early as possible.

....

The above press note was laid on the table of Indian Legislative Assembly on February 7, 1946.

....

APPENDIX–III

RECIPIENTS OF PBC AWARDS, 1998-1999

S.No.	Category/Field	Type of Award	Recipient
	Religion		
1.	*Qirat*	Super Star	Qari Obaidur Rahman
2.	*Qirat*	Special	Qari Fida Muhammad
3.	*Qirat*	Excellence	Qari Ghulam Rasool
4.	*Naat Khwani*	Super Star	Qari Waheed Zafar
5.	*Naat Khwani*	Super Star	Ms Muniba Shaikh
6.	*Naat Khwani*	Excellence	Umme Habiba
7.	*Naat Khwani*	Excellence	Siddique Ismail
	Presentation		
8.	Announcement	Super Star	Badar Rizwan
9.	Announcement	Excellence	Amirul Hasan Zaidi
10.	Compering	Excellence	Tauseeq Haider
	Music		
11.	*Qawwali*	Super Star	Munshi Raziuddin
12.	*Qawwali*	Excellence	Mehar Ali, Sher Ali
13.	Classical Vocal	Super Star	Ustad Ghulam Husain Shaggan
14.	Classical Flute	Super Star	Ustad Salamat Husain
15.	Classical Sarangi	Super Star	Ustad Allah Rakha Khan
16.	Light Vocal	Super Star	Mehdi Hasan
17.	Light Vocal	Super Star	Ghulam Ali
18.	Light Vocal	Super Star	Iqbal Bano
19.	Light Vocal	Super Star	Malika Pukhraj
20.	Light Vocal	Super Star	Farida Khanum
21.	Light Folk	Super Star	Reshman
22.	Light Folk	Super Star	Pathane Khan
23.	Light Folk	Super Star	Allan Faqir
24.	Light Folk	Super Star	Zar Sange
25.	Light Composer	Super Star	Ustad Nazar Hussain
26.	Light Vocal	Excellence	Waris Baig

#	Category	Award	Name
27.	Light Vocal	Excellence	Humaira Arshad
28.	Light Composer	Excellence	Mujahid Hussain
	Drama		
29.	Acting	Super Star	Qazi Wajid
30.	Acting	Super Star	Mahmood Ali
31.	Acting	Super Star	Talat Husain
32.	Acting	Super Star	Irfan Khusat
33.	Acting	Super Star	Begum Khurshid Shahid
34.	Acting	Super Star	Fatima Khanum
35.	Playwright	Super Star	Bano Qudsia
36.	Playwright	Super Star	Basit Saleem
37.	Playwright	Excellence	Mohsin Usmani
38.	Acting	Excellence	Firdaus Jamal
39.	Acting	Excellence	Nargis Rashid
40.	Acting	Excellence	Nazir Husaini
41.	Feature writing	Excellence	Prof. Shabbir Ahmed Qadri
	News		
42.	Newsreading (English)	Super Star	Jahanara Sayeed
43.	Newsreading (English)	Super Star	Prof. Anita Ghulamali
44.	Newsreading (English)	Super Star	Shaista Zaid
45.	Newsreading (Urdu)	Super Star	Ms. Shamim Ejaz
46.	Newsreading (Urdu)	Excellence	Ishrat Saqib
47.	Newsreading (English)	Excellence	Marium Gilani
48.	Editing	Excellence	Faqir Ahmad
49.	Reporting	Excellence	Abdul Hameed
50.	Regional Reporting	Excellence	Gulzar Ahmed
51.	Translation	Excellence	Muhammad Hayat
52.	Monitoring	Excellence	Reaz Ahmed
53.	Newsreading (Punjabi)	Excellence	Ghias Malik
54.	Newsreading (Sindhi)	Excellence	Ejaz Ahmed Chatha
55.	Newsreading (Pushto)	Excellence	Mushtaq-ur-Rehman
56.	Newsreading (Balochi)	Excellence	Abdus Samad Bangalzai

Sports Commentary

57.	Cricket (English)	Super Star	Omar Koreishi
58.	Cricket (English)	Super Star	Jamshed Marker
59.	Cricket (Urdu)	Super Star	Munir Husain
60.	Hockey	Super Star	S.M. Naqi
61.	Cricket (English)	Excellence	Chishti Mujahid
62.	Cricket (Urdu)	Excellence	Mirza Iqbal Baig
63.	Hockey (English)	Excellence	Amjad Aziz Malik

Production

64.	Religious	Excellence	Hifzur Rahman
65.	Music(CPU)	Excellence	Shahwar Hyder
66.	Drama	Excellence	Shama Khalid
67.	Disc Jockey Prog	Excellence	Rabia Akram
68.	Feature	Excellence	Arifullah
69.	Outside Broadcast	Excellence	Naveed Ahmad Chauhan
70.	Feature (Punjabi Darbar)	Excellence	Reaz Mahmood
71.	Documentary	Excellence	Fakhar Abbas
72.	Special Audience Programme	Excellence	Sajid Hasan Durrani
73.	Sports	Excellence	Muhammad Tasleem
74.	Language (Turkish)	Excellence	Ayfar Baseer
75.	Feature 'Aaina' and Compering	Excellence	Yasmin Jamil

Engineering

76.	Equipment Production	Excellence	Naeem Akhtar
77.	Transmitter	Excellence	Syed Moenuddin
78.	Recording	Excellence	Mahmood Ahmed
79.	Outside Broadcast	Excellence	Muhammad Aslam
80.	Maintenance	Excellence	Muhammad Ishaq
81.	Project	Excellence	Gul Taj
82.	Innovation	Excellence	Imtiazuddin
83.	Broadcasting House	Excellence	Noorul Haq Rashdi
84.	Mechanic Diesel	Excellence	Sher Ali

Finance & Accounts

85.	Accounts Headquarters	Excellence	Akhtar Husain Shaikh
86.	Accounts Units	Excellence	Israr Ahmed Bhatti
87.	Accounts Sales	Excellence	Khalid Farooq

Administration
#			
88.	Supporting Staff	Excellence	Raana Shamim
89.	Mali (Gardener)	Excellence	Shaukat Ali Bilal
90.	Driver	Excellence	Muhammad Ashiq Butt
91.	Qasid	Excellence	Maqsood Anwar
92.	Security Guard	Excellence	Abdul Aziz
93.	Sanitary Worker	Excellence	Yousuf Masih

Marketing
#			
94.	Friendly Client	Excellence	Lever Brothers (Saleem Durrani)
95.	Friendly Advertising Agency	Excellence	R. Lintas (Abid Ali Khan)
96.	Salesman	Excellence	Mashood Ahsan
97.	Drama/Compering	Lifetime Achievement	Mohni Hameed
98.	Playwright	Lifetime Achievement	Ashfaq Ahmed
99.	Music Vocal	Lifetime Achievement	Noor Jahan
100.	Broadcasting	Special	Bashir Ahmad Baloch
101.	Broadcasting	Special	Ilyas Ishaqie
102.	Broadcasting	Special	Masud Qureshi
103.	Broadcasting	Special	Najmul Hasnain
104.	Broadcasting	Special	Sikandar Baloch
105.	Broadcasting	Special	Yawar Mehdi

Appendix–IV

Recipients of PBC Awards, 1999-2000

S.No.	Category/field	Type of Award	Recipient
	Religion		
1.	Qari	Excellence	Buzurg Shah
2.	Naat Khwan	Excellence	Marghoob A. Hamdani
	Drama/Feature		
3.	Writer	Excellence	Prof. Rab Nawaz Moonis
4.	Artist (Male)	Excellence	Shamshad Ali Khan
5.	Artist (Female)	Excellence	Nasreen Aftab
6.	Stock Character	Special	Shamsher Haider
7.	Stock Character	Special	Saleh Muhammad Shah
8.	Stock Character	Special	Said Rehman Shino
	Music		
9.	Classical Vocal	Super Star	Ustad Fateh Ali Khan
10.	Classical Instrumental	Super Star	M. Subhan Rathor
11.	Classical Vocal (Male)	Excellence	Badaruzzaman Qamaruzzaman
12.	Light Vocal Male	Excellence	Ijaz Qaiser
13.	Light Vocal Female	Excellence	Mehnaz
14.	Composer	Excellence	Ustad Raees Khan
15.	Instrumental	Special	Taj Muhammad Tajil
16.	Qawwal	Excellence	Zaki Taji
	Presentation		
17.	Announcer	Excellence	Farzana Irum
18.	Compere Disc Jockey	Excellence	S. M. Saleem
19.	Compere English Programme	Excellence	Mazhar Nisar
20.	Announcer	Special	Mushabbar Zaidi
21.	Announcer	Special	Shakira Siddiqui

22.	DJ FM 101 English	Excellence	Kashif Khan
23.	DJ FM 101 Urdu	Excellence	Sufian J. Yusufi

Sports Commentary

24.	Hockey	Excellence	Amjad Aziz Malik
25.	Cricket (Urdu)	Excellence	Ehtisham-ul-Haq
26.	Cricket (English)	Excellence	Pervez Asghar Mian
27.	Cricket (English)	Special	Shahzad Humayun

Production

28.	Religious Prog	Excellence	Hifz-ur-Rehman
29.	Subh-i-Pakistan	Excellence	Abdul Hafeez
30.	Drama/Feature	Excellence	Kaneez Fatima
31.	Current Affairs	Excellence	Mubashar Ahmed
32.	Special Audience Programme	Excellence	Iffat Jabbar
33.	Music	Excellence	Khalid Toor
34.	Documentary/ Investigative Report	Excellence	Arifullah
35.	English Prog	Excellence	Kanwal Naseer

News

36.	Newsreader (Urdu)	Excellence	Muhammad Umar
37.	Newsreader (English)	Excellence	Riaz Ahmed Khan
38.	Newsreader (Regional)		Excellence Bilqees Haider Sindhi
39.	News Reporter	Excellence	Pervez Akhtar Zia
40.	News Editor	Excellence	Muhammad Niaz
41.	Editor Language	Excellence	Mukhtar Ahmad Khaleel

Engineering

42.	Drama Recording	Excellence	S.A. Sami
43.	Music Recording	Excellence	Zaheer Ahmed Ayaz
44.	Writer, Poet	Lifetime Achievement	Ahmed Nadeem Qasmi
45.	Music	Lifetime Achievement	Mehdi Hasan
46.	Music	PBC Noor Jahan	Tarannum Naz Laila
47.	Broadcasting	Best Station	Quetta

TABLE - 1
RADIO PAKISTAN AT A GLANCE

	August 1947 Pakistan	West Pakistan	December 1972	December 1996	June 2000
Number of Stations	3	2	9	23	24
Stations with Second Channel	-	-	2	5	5
FM Stations	-	-	-	3	3
Programme Hours (Home Service)	27	18	137	290	365
Coverage: Area %	7	6	33	75	78
Population %	16	21	78	95	96
No. of Languages in Home Service	6	4	17	21	21
No. of Bulletins	19	12	71	108	108
Transmitter Power: KW	20	15	856	3,844	3,854
MW KW	20	15	396	2,691	2,701
SW KW	-	-	460	1,141	1,141
FM KW	-	-	-	12	12
No. of Transmitters	3	2	21	45	46
MW	3	2	10	26	27
SW	-	-	11	14	14
FM	-	-	-	5	5
External Services					
Duration (Hours)	-	-	16	20	12
Languages	-	-	15	15	17
World Service (Hours)	-	-	-	10	10
Commercial Service (Hours Daily)	-	-	16	60	60
CPU Material in Archives (000 minutes)	-	-	150	500	1013
No. of Employees	484	NA	2,512	5,349	5,350
No. of Units/Offices	7	5	32	54	55

TABLE - 2
DIRECTORS GENERAL
RADIO PAKISTAN

#	Name	From		To
1.	Mr Z. A. Bokhari	06.08.1947	to	03.06.1959
2.	Mr Rashid Ahmad	04.06.1959	to	09.09.1962
3.	Mr Zahur Azar	10.09.1962	to	31.12.1965
4.	Syed Munir Husain	11.01.1966	to	15.08.1969
5.	Mr Mafizur Rahman.	14.11.1969	to	15.04.1971
6.	Syed Ijlal Haider Zaidi	15.04.1971	to	04.02.1972
7.	Khwaja Shahid Hosain	04.02.1972	to	19.12.1972

PAKISTAN BROADCASTING CORPORATION

#	Name	From		To
	Khwaja Shahid Hosain	20.12.1972	to	12.06.1974
8.	Syed Ijlal Haider Zaidi	12.06.1974	to	26.11.1977
9.	Mr Q.A. Saeed	26.11.1977	to	22.09.1984
10.	Mr K.G. Murtaza	11.10.1984	to	25.03.1986
11.	Mr Saleem A. Gilani	12.05.1986	to	10.12.1988
12.	Mr Agha Nasir	10.12.1988	to	05.09.1990
13.	Mr Saleem A. Gilani	05.09.1990	to	04.09.1991
14.	Mr Agha Nasir	05.09.1991	to	25.07.1992
15.	Mr Abdul Khaliq Awan	25.07.1992	to	28.06.1995
16.	Mr Mohammad Abbas	26.09.1995	to	14.05.1997
17.	Khwaja Ijaz Sarwar	15.05.1997	to	10.08.1998
18.	Mr S. Anwar Mahmood	10.08.1998	to	13.10.2000
19.	Mr Salim Gul Shaikh	13.10.2000		

TABLE - 3
EXTERNAL SERVICES
(UP TO 13.08.1997)

S. No.	Name of Service	Date of Introduction	Timings(PST) UTC in Brackets.	Duration (Hours)	Target Areas/ Countries
1.	Dari	14.08.1949 (04.12.1939)	2100 – 2300 (1600 – 1800)	2	Afghanistan.
2.	Arabic	14.08.1949	2200 – 2400 (1700 – 1900)	2	Saudi Arabia, Qatar, UAE, Bahrain, Oman, Kuwait, Dubai, Iraq, Syria, Jordan, Lebanon, Yemen, Egypt, Libya, Algeria, Sudan, Chad, Niger, Mauritania, Tunisia, Morocco.
3.	Irani	14.08.1949	0830 –1030 (0330 – 0530)	2	Iran
4.	Myanmar	14.08.1949	1200 – 1300 (0700 – 0800)	1	Myanmar. Discontinued from 04.05.1997
5.	Gujarati	10.03.1952	0815 – 0915 (0315 – 0415)	1	Ethiopia, Uganda, Kenya, Somalia, Tanzania, Zaire, Angola, Zambia, Zimbabwe, Botswana, Namibia, South Africa.
6.	Hindi	01.07.1952	0530 – 0630 (0030 – 0130) 1700 – 1800 (1200 – 1300)	2	India (North)
7.	Turki	01.09.1957	1930 – 2030 (1430 – 1530)	1	Afghanistan and Central Asia.
8.	Turkish	01.09.1957	2200 – 2300 (1700 – 1800)	1	Turkey.
9.	Indonesian	01.02.1964	1400 – 1500 (0900 – 1000)	1	Indonesia, Malaysia and Brunei Darussalam.
10.	Tamil	16.09.1965	1530 – 1630 (1030 – 1130)	1	India (South) and Sri Lanka.
11.	Swahili	16.02.1966	1145 – 1245 (0645 – 0745)	1	Kenya, Mauritius, East Africa, Tanzania, Ethiopia, Zimbabwe. Discontinued from 14.08.1997

12.	Nepali	10.1970	1820 – 1920 (1320 – 1420)	1	Nepal. The Service was originated from Dhaka(East Pakistan) and remained there after dismemberment of Pakistan in December 1971.
13.	Sylheti	01.10.1971	1835 – 1850 (1335 – 1350)	15 mts.	Assam (India), Sylhet (Bangladesh), UK. Later part of Mitali Service. Discontinued from 14.08.1997
14.	French	01.11.1971	0030 – 0130 (1930 – 2030)	1	France, Algeria, Senegal, Gambia, Mauritania, Ghana, Nigeria, Tunisia, Mali, Guinea, Sierra Leone, Niger and Chad.
15.	Mitali (Bangla)	14.04.1972	0600 – 0700 (0100 – 0200) (1700 – 1900) (1200 – 1400)	3	Bangladesh, West Bengal (India), Assam (India), Myanmar, Bangkok (Thailand).
16.	Hazargi	01.03.1975	1900 – 2000 (1400 – 1500)	1	Afghanistan.
17.	Chinese	12.05.1997	1700 – 1730 (1200 – 1230)	30 mts.	China.
18.	Russian	12.05.1997	2000 – 2030 (1500 – 1530)	30 mts.	Central Asian Countries.

TABLE – 4
EXTERNAL SERVICES
(Effective 14.08.1997)

S. No.	Name of Service.	Timings(PST) UTC in brackets.	Duration (Minutes)	Target Areas/ Countries.
1.	French	0030 – 0100 (1930 – 2000)	30	France, Algeria, Senegal, Mauritania, Ghana, Nigeria, Chad.
2.	Assami* (English)	0545 – 0615 (0045 – 0115)	30	North East India, UK.
3.	Bangla – I @	0615 – 0700 (0115 – 0200)	45	Bangladesh, SAARC.
4.	Hindi – I @	0700 – 0745 (0200 – 0245)	45	India.
5.	Tamil – I @	0745 – 0815 (0245 – 0315)	30	South India, Sri Lanka.
6.	Gujarati	0830 – 0900 (0330 – 0400)	30	Ethiopia, Uganda, Tanzania, Somalia, Angola, Zimbabwe, South Africa.
7.	Indonesia	1400 – 1430 (0900 – 0930)	30	Indonesia, Malaysia, Brunei Darussalam.
8.	Tamil- II @	1530 – 1600 (1030 – 1100)	30	South India, Sri Lanka.
9.	Hindi-II @	1600 – 1700 (1100 – 1200)	60	India.
10.	Sinhali **	1645 – 1715 (1145 – 1215)	30	Sri Lanka.
11.	Bangla – II @	1700 – 1730 (1200 –1230)	30	Bangladesh, SAARC
12.	Chinese	1700 – 1730 (1200 – 1230).	30	China
13.	Nepali**	1715 – 1745 (1215-1245)	30	Nepal.
14.	Irani	1800 – 1845 (1300 – 1345)	45	Iran.
15.	Hazargi	1900 – 1930 (1400 – 1430)	30	Afghanistan.
16.	Turki	1930 – 2000 (1430 –1500)	30	Afghanistan, Central Asia.
17.	Russian	2015 – 2045 (1515 – 1545)	30	Central Asian countries.
18.	Dari	2100 – 2200 (1600 – 1700)	60	Afghanistan

19.	Turkish	2200 – 2230 (1700 – 1730)	30	Turkey.
20.	Arabic	2300 – 2345 (1800 – 1845)	45	Saudi Arabia, Qatar, UAE, Bahrain, Iraq, Egypt, Yemen, Libya, Sudan, Chad, Lebanon, Kuwait.
	17		**720**	

* = Assami Service introduced on 01.01.1999.
@ = Duration raised and broadcast in two chunks from 01.01.1999.
** = Sinhali and Nepali Services introduced on 26.03.2000.

TABLE – 5
WORLD SERVICE
(Date of Commencement: April 21, 1973)

S. No.	Name of Service	Timings (PST) UTC in brackets.	Duration	Target Areas/Countries.
1.	South East Asia	0600 – 0700 (0100 – 0200)	1 hour	India, Bangladesh, Indonesia, Myanmar, Malaysia, Brunei Darussalam, Australia, Philippines, Vietnam, Singapore, Thailand, Nepal, New Zealnad and Laos.
2.	Gulf and Middle East	1000 – 1200 (0500 – 0700)	2 hours	Afghanistan, Iran, Iraq, Kuwait, Qatar, Oman, UAE, Bahrain, Syria, Turkey, Jordan, Saudi Arabia, Egypt, Yemen, Nigeria, Sudan, Tunisia, Libya and Muscat.
3.	Western Europe & UK	1300 – 1600 (0800 – 1100)	3 hours	England, Finland, Netherlands, Sweden, Norway, Denmark, Italy, Germany, France, Poland, Austria, Hungary, Spain, Ireland, Switzerland, Yugoslavia, Turkey, Czech, Romania, Portugal, Belgium and CIS.
4.	Gulf and Middle East	1830 – 2030 (1330 – 1530)	2 hours	As at S.No. 2
5.	Western Europe & UK	2200 – 2400 (1700 – 1900)	2 hours	As at S.No. 3

Total Duration = **10 hours.**
Languages = **Urdu and English.**

TABLE – 6
NATIONAL NEWS BULLETINS

S. No.	TIME	LANGUAGE	DURATION (MINUTES)
1.	0600 hrs.	Urdu	03
2.	*0700 hrs.	Urdu	10
3.	*0800 hrs.	English	10
4.	0900 hrs.	Urdu	03
5.	1000 hrs.	Urdu	03 (Sunday only)
6.	1100 hrs.	Urdu	04
7.	1200 hrs.	Urdu	03
8.	1204 hrs.	Punjabi	04
9.	*1300 hrs.	English	03
10.	1304 hrs.	Sindhi	04
11.	1400 hrs.	Urdu	03
12.	1404 hrs.	Pushto	04
13.	*1500 hrs.	Urdu	05
14.	1600 hrs.	English	04
15.	1604 hrs.	Balochi	04
16.	1700 hrs.	Urdu	05
17.	*1800 hrs.	English	10
18.	1900 hrs.	Urdu.	03
19.	*2000 hrs.	Urdu (*Khabarnama*)	30
20.	*2100 hrs.	English	10
21.	2200 hrs.	Urdu	10
22.	2300 hrs.	Urdu	02
22		**6 Languages**	**137**

* = These seven bulletins in Urdu and English with total duration of 78 minutes are put on Internet also.
Wesbsite: www.radio.gov.pk.

TABLE – 7
SUMMARY OF NEWS BULLETINS

S. No.	Types of Bulletin	As on 13.08.1997		As on 30.06.2000	
		No. of Bulletins	Duration (Minutes)	No. of Bulletins.	Duration (Minutes)
1.	National	19	114	22	137
2.	Regional	25	169	17	142
3.	FM	2	4	7	21
4.	External	16	102	20	100
5.	GOS	3	45	*	-
6.	Local	41	201	42	206
7.	Commercial	1	5	@	-
8.	Arabic	1	10	#	-
9.	News Summary	1	5	*	-
	Total	**109**	**655**	**108**	**606**

@ = Included in *Khabarnama*.
\# = Included in External Service (Arabic)
* = Dropped as some bulletins are available on Internet.

TABLE - 8
BROADCAST CENTRES (RADIO STATIONS) IN PAKISTAN

Radio Station	Power of T/R (KW)	Type of T/R	Frequency (kHz)	Date of operation of T/R	No. of Studios	Date of Operation of studios	Remarks
01. Islamabad	1000	MW	585	14.08.1977	20	07.05.1977	Large variety of
	2x250	SW					equipment of
	2x100	SW					RCA Philips
	2x100	SW					Ampex, Denor
	1x10	SW					PBC, AWA, NEC
	1x2	FM	101 mHz.				etc. is being used.
02. Lahore	100	MW	630	15.11.1964	8	09.08.1965	5 KWMW in
	50	MW	1080	15.10.1975			operation on
	2	FM	101 mHz				16.12.1937
03. Karachi	100	MW	825	23.03.1977	10	16.07.1951	100 watt MW
	10	MW	639	14.08.1949			T/R in operation
	5	FM	101 mHz				on 14.08.1949
	2 x 50	SW		14.08.1949			

#	Station	Power	Type	Freq	Date	Col6	Date2	Notes
04.	Quetta	150	MW	756	25.04.1974	10	01.06.1961	1 KWMW T/R in operation on 17.10.1956
		10	MW	855	01.06.1961			
		10	SW		17.10.1962			
05.	Peshawar	300	MW	540	22.04.1977	6	28.04.1985	10 KMW T/R operation in 1942
		100	MW	1377	13.11.1993			
		10	SW		16.01.1960			
06.	Rawalpindi	10	MW	1152	21.05.1953	6	21.05.1953	100 watt MW T/R operative on 01.09.1950
		10	SW		21.05.1953			
					21.05.1953			
07.	Multan	120	MW	1035	21.11.1970	6	21.11.1970	
		2	FM					
08.	Bahawalpur	10	MW	1341	01.07.1975	3	01.07.1975	
09.	Faisalabad	10	MW	1476	01.10.1989	3	01.10.1989	250 watt MW T/R operative on 15.09.1982
10.	Hyderabad	120	MW	1008	04.02.1971	6	1970-71	1 KWMW T/R in operation on 17.08.1955
		10	MW	1098	25.12.1959			
11.	Khairpur	100	MW	927	03.03.1981	4	03.05.1986	
12.	D.I.Khan	10	MW	1404	15.01.1981	2	15.01.1981	
13.	Khuzdar	300	MW	567	03.09.1991	4	30.06.1988	250 watt MW T/R operative on 17.06.1981
14.	Sibi	0.25	MW	1584	01.10.1989	1	01.10.1989	
15.	Gilgit	10	MW	1512	01.10.1989	2	02.04.1979	250 watt MW T/R operative on 02.04.1979
16.	Turbat	0.25	MW	1584	04.01.1981	1	04.01.1981	
17.	Skardu	10	MW	1557	14.12.1983	2	16.04.1979	250 watt MWT/R operative on 16.04.1979
18.	Abbottabad	0.25	MW	1602	01.10.1989	1	01.10.1989	
19.	Chitral.	0.25	MW	1584	18.08.1993	1	18.08.1993	
20.	Larkana	1	FM	101 mHz	26.09.1995	1	26.09.1995	
21.	Rawalpindi-II	150	MW	792	02.10.1983	4	15.10.1960	
		1	SW		15.10.1960			
22.	Rawalpindi-III	100	SW		07.08.1968	3	21.05.1953	500 watt SW T/R operative on 18.04.1948
23.	Loralai	10	MW	1251	22.07.1996	-		
24.	Zhob	10	MW	1449	21.04.1998	4	-	

T/R = Transmitter.

TABLE – 9
FIXED TRANSMISSION SCHEDULE

Radio Stations	M.W. Channel - I (PST)		M.W. Channel - II (PST)	
	Ist Transmission	2nd Transmission	Ist Transmission	2nd Transmission
Islamabad	0545 – 1105	1300 – 2400		2100 – 2300
		1500 – 2400 (Fri)	-	
Rawalpindi	0545 – 0905	1100 – 2308	-	-
		0545 – 1320 (Fri)	1500 – 2308 (Fri)	
Lahore	0545 – 0905	1100 - 2400		
	0545 – 1320 (Fri)	1500 – 2400 (Fri)	0900 – 1108	1730 – 1930
Multan	0445 – 0905	1100 - 2310	-	-
	0545 – 1320 (Fri)	1500 – 2310 (Fri)		
Bahawalpur	-	1350 – 2310	-	-
Faisalabad	0545 – 17 15	-	-	-
Karachi	0545 – 0905	1100 – 2400	0815–1145	-
	0545 – 1320 (Fri)	1500 – 2400 (Fri)		
Hyderabad	0545 – 0904	1100 - 2308	0900 – 1205	-
	0545 – 1320 (Fri)	1500 – 2308 (Fri)		
Khairpur	0545 – 0905	1100 - 2310	-	-
	0545 – 1320 (Fri)	1500 – 2310 (Fri)		
Larkana	0850 – 1510	-	-	-
	0850 – 1320 (Fri)			
Peshawar	0545 – 0904	1100 – 2308		
	0545 – 1320 (Fri)	1500 – 2308 (Fri)	-	1600 – 1804
D.I. Khan	-	1400 – 2100	-	-
Abbottabad	-	1345 – 1915		
Chitral	-	1550 – 2015	-	-
Quetta	0545 – 1304	1500 – 2310	0700 –0904	1100 – 2310
Khuzdar	-	1655 – 2310	-	-
Turbat	-	1800 – 2310	-	-
Sibi	-	1255 – 1608	-	-
Skardu	-	1500 – 2200	-	-
Gilgit	-	1400 – 2200	-	-
Loralai	-	1700 – 2100	-	-
Zhob	-	1645 – 2100	-	-
Rawalpindi-II	0545 – 0950	1500 – 2315	-	-
Rawalpindi-III	0545 – 0935	1300 – 1710	-	-

TABLE - 10
STAFF POSITION

S. No.	Scale No.	As on 31. 12. 1996			As on 30. 06. 2000		
		Sanctioned Strength	Actual Strength	Vacant	Sanctioned Strength	Actual Strength	Vacant
1.	1	1,397	1,092	305	1,396	830	566
2.	1-A	128	94	34	129	95	34
3.	2	718	610	108	663	481	182
4.	3	972	762	210	960	644	316
5.	4	565	475	90	538	392	146
6.	4-A	84	68	16	84	62	22
7.	5	543	281	262	625	267	358
8.	6	625	460	165	546	370	176
9.	7	182	165	17	247	206	41
10.	8	98	87	11	120	111	09
11.	9	31	31	-	36	33	3
12.	M-III	5	3	2	5	3	2
13.	M-II	1	1	-	1	1	-
	Sub Total	5,349	4,129	1,220	5,350	3,494	1,856
	Staff Artists	467	170	297	463	143	320
	Total	**5,816**	**4,299**	**1,517**	**5,813**	**3,637**	**2,176**

TABLE – 11
B.R. LICENCES – ISSUED/RENEWED

YEAR	NO. OF LICENCE. (000)
1960-61	314
1961-62	420
1962-63	447
1963-64	480
1964-65	562
1965-66	815
1966-67	725
1967-68	701
1968-69	1139
1969-70	1100
1970-71	968
1971-72	1039
1972-73	1572
1973-74	1493
1974-75	1388
1975-76	1521
1976-77	1229
1977-78	1604
1978-79	1489
1979-80	1800
1980-81	1528
1981-82	1336
1982-83	1328
1983-84	1252
1984-85	1208
1985-86	1140
1986-87	1259
1987-88	1107
1988-89	945
1989-90	1078
1990-91	1309
1991-92	830
1992-93	743
1993-94	698
1994-95	589
1995-96	523
1996-97	474
1997-98	442
1998-99	423

TABLE – 12
BROADCAST RECEIVER LICENCE COLLECTIONS

Year	Gross Collection (000 Rs.)	PPO Share (000 Rs)	%	PBC Share (000 Rs.)	%
1972-73	13195	2239	17.0	10956	83.0
1973-74	23430	2460	10.5	20970	89.5
1974-75	22717	2638	11.6	20079	88.4
1975-76	31926	2883	9.3	29043	90.7
1976-77	27583	2925	10.6	24658	89.4
1977-78	56713	4024	11.0	32689	89.0
1978-79	34500	5499	15.9	29001	84.1
1979-80	45000	6170	13.6	38830	86.4
1980-81	38196	6330	16.6	31866	83.4
1981-82	36505	6413	17.6	30092	82.4
1982-83	37,824	8088	21.4	29736	78.6
1983-84	36001	9650	26.8	26351	73.2
1984-85	36022	10117	28.1	25905	71.9
1985-86	36249	9710	26.8	26539	73.2
1986-87	33759	10753	31.9	23006	68.1
1987-88	35846	12803	36.1	23043	63.9
1988-89	34297	13612	36.8	20685	63.2
1989-90	33014	14603	44.2	18411	55.8
1990-91	44465	14702	33.1	29763	66.9
1991-92	45246	15992	35.3	29254	64.7
1992-93	44366	17310	39.0	27056	61.0
1993-94	42291	17323	45.7	24968	54.3
1994-95	39073	20237	51.8	18836	48.2
1995-96	37060	22739	61.4	14321	38.6
1996-97	33635	18270	54.3	15365	45.7
1997-98	31698	19399	61.2	12299	38.8
1998-99	28656	18798	65.6	9858	34.4

TABLE - 13
TOTAL INCOME AND EXPENDITURE

(In million Rupees)

Year	BR Licence Fee	Advertisement Income	Govt. Grant	Misc. Income	Total Income	Total Expenditure	Surplus (Deficit)	Dev. Grant
1972-73	10.95	4.69	11.19	0.17	27.01	24.06	2.95	14.60
1973-74	20.96	10.20	26.06	0.79	58.02	56.36	1.66	31.44
1974-75	20.07	14.02	34.61	3.62	72.33	79,01	(6.68)	58.74
1975-76	29.02	14.88	51.03	2.29	97.24	94.43	2.81	48.11
1976-77	24.65	13.71	74.32	8.58	121.27	118.64	2.63	40.41
1977-78	32.68	15.14	83.13	1.63	132.58	126.54	6.04	47.90
1978-79	28.99	14.81	99.36	1.28	144.45	133.91	10.54	28.05
1979-80	38.82	17.16	100.57	1.18	157.74	141.53	16.21	26.50
1980-81	31.86	25.09	109.18	1.87	168.00	152.07	15.93	26.21
1981-82	30.09	28.21	120.57	2.95	181.82	175.32	6.50	34.14
1982-83	29.73	32.90	138.66	0.58	201.88	202.19	(0.31)	38.87
1983-84	26.35	34.04	188.55	0.40	253.34	250.89	2.45	33.94
1984-85	25.90	40.43	189.60	0.92	256.86	260.76	(3.90)	29.42
1985-86	26.54	48.11	236.93	0.51	312.10	301.69	10.41	36.13
1986-87	23.00	51.68	248.77	1.04	324.50	324.28	0.22	45.22
1987-88	23.04	52.05	296.49	2.30	373.89	364.14	9.75	52.00
1988-89	20.69	50.40	275.55	2.12	348.76	379.26	(30.50)	44.17
1989-90	18.41	57.34	285.58	2.11	363.44	408.44	(45.00)	45.83
1990-91	29.76	44.98	337.54	2.39	414.67	433.26	(18.58)	48.97
1991-92	29.25	45.63	432.23	2.82	509.94	508.41	1.53	13.63
1992-93	27.05	57.76	400.68	3.62	489.11	530.38	(41.27)	68.33
1993-94	24.97	53.24	423.23	3.60	505.04	571.15	(66.11)	34.15
1994-95	18.84	68.66	547.90	7.91	643.31	662.86	(19.55)	78.18
1995-96	14.32	74.90	568.40	13.67	679.29	733.07	(61.78)	150.00
1996-97	15.36	74.72	555.43	14.23	659.74	737.88	(78.14)	37.07
1997-98	12.30	93.03	588.37	11.07	704.77	771.20	(66.43)	35.35
1998-99	9.86	86.11	564.98	7.18	668.13	797.99	(129.85)	15.47
1999-2000	-	90.36	761.99	21.21	873.56	889.20	(15.64)	70.00

TABLE - 14
MATERIAL AVAILABLE IN NATIONAL SOUND ARCHIVES
(As on 30.06.2000)

Recorded Material (Total)		10,13,451 minutes
Tilawat-i-Kalam-i-Pak		30,000 minutes
Hamd-o-Naat		5,550 minutes
Speeches of National Leaders (Governors-General, Presidents, Prime Ministers)		8,15,000 minutes
Interviews of leaders and workers of Pakistan Movement		3,082 minutes
Music		1,10,744 minutes
Vocal and Instrumental	35,550 minutes	
Songs in Regional Languages	14,650 minutes	
Folk Songs	6,200 minutes	
Folk Instrumental Music	3,200 minutes	
Classical Instrumental	16,884 minutes	
Classical Vocal	15,200 minutes	
Light Classical Vocal	4,540 minutes	
Patriotic Songs	6,700 minutes	
Theatre Music	1,620 minutes	
Qawwali	6,200 minutes	
Tafseer-o-Taleemul Quran (Urdu)		6050 minutes
Tafseer-o-Taleemul Quran (Kashmiri)		7,500 minutes
Drama		33,165 minutes
Poetic Recitations by Poets.		2,360 minutes
No. of historians/intellectuals whose recordings are available		200
No. of poets, drama artists, narrators, writers		2,225
No. of Political Personalities		338
No. of *Qaris*		88
No. of foreign VIP's		282
No. of countries on CPU's regular mailing list		55

Chronology

CHRONOLOGY OF BROADCASTING EVENTS

1926	March	Indian Broadcasting Company (IBC) formed (Private).
1927	July 23,	Bombay Station of IBC opened, beginning organized broadcasting in Indo-Pakistan subcontinent.
	August 26,	Calcutta Station of IBC opened.
1928	——	A small transmitting station set up at Lahore by YMCA.
1930	April 1,	Broadcasting placed under direct control of Government under the title 'Indian State Broadcasting Service'(ISBS).
1934	January 1,	Indian Wireless Telegraphy Act (1933) comes into force.
1935	January,	NWFP Govt. sets up a 250 watt transmitting station at Peshawar for community listening.
	March 1,	Office of Controller of Broadcasting created under Department of Industries and Labour of the Government.
	August 30,	Mr.Lionel Fielden assumed charge as first Controller of Broadcasting.
1936	January 1,	Delhi Station opened.
	June 8,	Name of ISBS changed to All India Radio (AIR)
	July 9,	Mr.A.S.Bokhari, Station Director Delhi, becomes Deputy Controller of Broadcasting.
	July 16,	Peshawar Station inaugurated.
1937	January 24,	First Station Directors' Conference held at Delhi.
	April 1,	Peshawar Station taken over by Government of India from Government of NWFP.
	September 1,	Lahore Station of YMCA closed.
	December 16,	Lahore Station of AIR went on the air.
1939	March 1,	Peshawar Station converted into a relay station.
	September,	Centralization of news bulletins in all languages at Delhi.
	November 12,	Quaid-i-Azam's first radio broadcast from Bombay Station on Eid Day.

	December 16,	Dhaka Station Opened.
1941	October 24,	Department of Information and Broadcasting set up.
1942	July 16,	Shifting of Peshawar Station to a regular Broadcasting House and formal inauguration.
1943	February,	Designation of Controller of Broadcasting changed to Director General AIR.
1947	June 3,	Broadcasts by Lord Mountbatten, Pandit Jawaharlal Nehru, Quaid-i-Azam Mohammad Ali Jinnah and Sardar Baldev Singh regarding partition of Indian subcontinent.
1947	August 14,	Pakistan and Pakistan Broadcasting Service come into being.
1948	April 16,	Inauguration of Rawalpindi-III Station and 500 watt SW transmitter.
	August 14,	Inauguration of emergency Karachi Station and 100 watt MW transmitter.
	August 14,	First issue of fortnightly 'Ahang' published.
	August 25,	First issue of fortnightly 'Pakistan Calling' published.
	November 1,	Commissioning of 10 KW MW Transmitter at Landhi, Karachi.
1949	January 16,	Commissioning of first 7.5 KW SW Transmitter at Dhaka.
	May, 25,	Commissioning of 1 KW SW Transmitter at Lahore.
	August 14,	Commissioning of 50 KW SW Transmitter at Landhi, Karachi.
	August 14,	Arabic, Iranian, Afghan-Persian (Dari) and Burmese Services introduced.
	December 25,	Commissioning of 2nd 50 KW SW Transmitter at Landhi, Karachi.
1950	September 1,	Inauguration of Rawalpindi Station and 100 watt MW Transmitter.
1951	July 16,	Inauguration of new Broadcasting House at Karachi.
	December,	Emergency Monitoring Centre set up at Clifton, Karachi.
1952	January 18,	South East Asian, South Asian, Indonesian, Turkish and UK Services started on experimental basis.
	March 10,	Gujrati (East and South East African) Service introduced.
	July 1,	Hindi Service introduced.
	September 18,	South East Asian, South Asian, Indonesian, Turkish and UK Services made regular.
1953	May 21,	Commissioning of 10 KW MW Transmitter at Rawalpindi, replacing 100 watt MW Transmitter.

	December 25,	Commissioning of 10 KW SW Transmitter-I at Landhi, Karachi.
1954	March,	Commissioning of 10 KW SW Transmitter-II at Landhi, Karachi.
	June 14,	Inauguration of Chittagong Relay Station and 1KW MW Transmitter.
	June 25,	Inauguration of Rajshahi Relay Station and 1 KW MW Transmitter.
1955	August 17,	Inauguration of Hyderabad Station and 1 KW MW Transmitter.
1956	March,	International Monitoring Centre at BH Karachi.
	October 17,	Inauguration of Quetta Station and 1 KW MW Transmitter.
1957	September 1,	Turki and Turkish Services introduced.
1959	September 1,	Commissioning of one 10 KW SW Transmitter at Dhaka.
	December 25,	Commissioning of one 10 KW MW Transmitter at Hyderabad, replacing the 1 KW MW Transmitter installed earlier.
1960	January 16,	Commissioning of one 10 KW SW Transmitter at Peshawar.
	February 8,	Broadcasting House, Dhaka.
	July,	Additional studios at Rawalpindi
	———	Receiving Centre at Karachi.
	October 15,	Inauguration of Rawalpindi-II Station with 1 KW SW Transmitter.
	December,	Receiving Centre at Peshawar.
1961	June 1,	Commissioning of 10 KW MW Transmitter at Quetta.
	———	Commissioning of 2 KW MW Transmitter at Sylhet.
	———	Receiving Centre at Quetta.
1962	———	Commissioning of 10 KW MW Transmitter at Chittagong.
	May,	Receiving Centre at Lahore.
	———	Commissioning of 10 KW MW Transmitter at Rajshahi.
	———	Quetta Station shifts to present Broadcasting House.
	October 17,	Commissioning of 10 KW SW Transmitter at Quetta.
	December,	Broadcasting House at Hyderabad.
1963	———	Receiving Centre at Dhaka.
	May 13,	Commissioning of 100 KW MW Transmitter at Savar, Dhaka.

	June 5,	Commissioning of 10 KW MW Transmitter at Quetta.
	—	Extension to Broadcasting House, Dhaka.
	October, 17,	Commissioning of 10 KW SW Transmitter at Quetta.
1964	February 1,	Indonesian Service introduced.
	July,	Receiving Centre Islamabad.
	—	Broadcasting House, Chittagong.
	—	Broadcasting House, Rajshahi.
1965	March,	Commissioning of 100 KW MW Transmitter at Lahore.
	August 9,	Lahore Station shifts to new Broadcasting House.
	September 16,	Tamil Service introduced.
	October 16,	Commissioning of 100 KWMW Transmitter at Lahore.
1966	February 16,	Swahili Service introduced.
	May 16,	AIR starts special Urdu Service for 9 hours for listeners in sub-continent.
1967	March 23,	Second channel starts at Lahore.
	—	Inauguration of 10 KW MW Rangpur Station.
1968	August 7,	Commissioning of one 100 KW SW and one 10 KW SW Transmitters, Islamabad.
	—	Commissioning of one 100 KW SW Transmitter, Rawalpindi.
	—	Commissioning of one 100 KW SW Transmitter, Dhaka.
1969	—	Commissioning of one 1 KW MW Transmitter, Karachi.
1970	—	Short wave aerial system, Islamabad.
	July,	Transcription Service studios at Lahore.
	July,	Staff Training School and Technical Training School start functioning at Islamabad.
	October,	Nepali Service started from Dhaka.
	November 21,	Inauguration of Multan Station and 120 KW MW Transmitter.
	December,	Inauguration of Khulna Station and 10 KW MW Transmitter.
1971	February 4,	Commissioning of 120 KW MW Transmitter, Hyderabad.
	November 1,	French Service introduced.
	November 8,	AIR introduced Sindhi language programme in External Services.
	October 1,	Sylheti Service introduced.

CHRONOLOGY

	December,	Mitali Service (Bangla) started in Home Service.
1972	April 14,	Mitali (Bangla) Service made part of External Services.
	April 27,	Foundation-stone laid of Broadcasting House, Islamabad.
	December 20,	Commissioning of two 250 KW SW Transmitters, Islamabad.
1973	April 21,	Inauguration of World Service.
	———	Merger of Staff Training School and Technical Training School.
1974	April 25,	Commissioning of one 150 KW MW Transmitter, Quetta.
	May,	AIR introduces Balochi Service of 30 minutes in External Services.
	May 27,	Foundation-stone laid of 100 KW Transmitter, Khairpur.
	July 15,	Commissioning of two 100 KW SW Transmitters, Islamabad.
1975	March 1,	Hazargi Service introduced.
	August 18,	Inauguration of Bahawalpur Station and 10 KW MW Transmitter.
	October 15,	Commissioning of one 50 KW MW Transmitter Lahore.
1976	April 29,	Commissioning of 10 KW MW Transmitter, Rawalpindi-II.
1977	March 23,	Commissioning of 100 KW MW Transmitter Karachi.
	May 7,	Inauguration of Islamabad Station and 1000 KW MW Transmitter.
1979	April 2,	Inauguration of Gilgit Station and 250 watt MW Transmitter.
	April 16,	Inauguration of Skardu Station and 250 watt MW Transmitter.
1980	August 25,	BR Licence fee on one and two band transistor sets abolished in India.
1981	January 4,	Inauguration of Turbat Station and 250 watt MW Transmitter.
	January 15,	Inauguration of D.I. Khan Station and 10 KW MW Transmitter.
	March 21,	Inauguration of Khairpur Station and 100 KW MW Transmitter.
	June 17,	Inauguration of Khuzdar Station and 250 watt MW Transmitter.

1982	September 15,	Inauguration of Faisalabad Station and 250 watt MW Transmitter.
	November 23,	Commissioning of 10 KW MW Transmitter, Skardu.
1983	October 2,	Commissioning of 150 KW MW Transmitter, Rawalpindi-II.
1984	August 15,	World Service started broadcasting from Islamabad.
1985	April 1,	Licence fee on radio and television sets abolished in India.
	April 28,	Peshawar Station shifts to new building, and 300 KW MW Transmitter commissioned.
1986	May 7,	Inauguration of new Broadcasting House, Khairpur.
1987	January 1,	New Concept of Broadcasting introduced.
	March,	Fortnightly 'Ahang' becomes a monthly.
	December 31,	Pakistan Post Office issues commemorative postage stamp of 80 paisa on completion of one year of "Radio Pakistan's New Concept of Broadcasting".
1988	March,	PBC wins Pakistan Advertising Association 'Medium of the Year Award' for 1987.
1989	August 24,	Foundation-stone laid of Loralai Station.
	October 1,	Commissioning of 10 KW MW Transmitter, Gilgit.
	-do-	Commissioning of 10 KW MW Transmitter, Faisalabad.
	-do-	Inauguration of Sibi Relay Station and 250 watt Transmitter.
	-do-	Inauguration of Abbottabad Relay Station and 250 watt Transmitter.
1991	February 20,	300 KW MW Transmitter starts functioning at Khuzdar.
	September 3,	Formal inauguration of 300 KW MW Transmitter, Khuzdar.
1992	July 20,	Test Transmission begins at Chitral.
1993	April 8,	FM Broadcasting started formally as 'FM Special'.
	August 18,	Inauguration of Chitral Station and 250 watt MW Transmitter.
	November 13,	Test Transmission of 100 KW MW Transmitter as Channel-II begins, Peshawar.
1994	July,	Satellite communication used for national hook-up programmes.
1995	September 24,	Foundation-stone laid of Larkana Broadcasting House.
	-do-	Inauguration of Larkana Station and 1 KW FM Transmitter.

1996	July 20,	Inauguration of Loralai Station and 10 KW MW Transmitter.
1997	May 4,	Myanmar (Burmese) Service discontinued.
	May 12,	Chinese Service introduced.
	-do-	Russian Service for the Central Asian Muslim States introduced.
	August 14,	Swahili and Sylheti Services discontinued.
	-do-	Timings and duration of almost all External Services changed.
	-do-	English and Urdu national news bulletins go on Internet.
1998	January 26,	Sautul Quran, separate five-hour channel, started from six Stations for tilawat, translation and Quranic teachings.
	April 21,	Zhob Station with 10 KW MW Transmitter inaugurated.
	May 19,	Golden Jubilee celebrations of Radio Pakistan/Pakistan Broadcasting Corporation, inaugurated by Prime Minister.
	October 1,	FM 101 Channel from Islamabad, Lahore and Karachi completely overhauled.
	November 2,	PBA FM School channel started for two hours.
	November 12,	Radio Pakistan programmes linked with Internet.
1999	January 1,	Assami Service introduced.
	July 1,	BRL fee abolished.
2000	February 10,	Duration of FM Service raised to 22 hours a day from Islamabad, Lahore and Karachi.
	March 26,	Nepali and Sinhali Services introduced.
2001	April 18,	News and Current Affairs Channel inaugurated.

BIBLIOGRAPHY

Abu News, Third Issue 1995, AIBD, Kuala Lumpur, Malaysia.
Baruah, U.L., *This is All India Radio*, Publications Division, Ministry of Information and Broadcasting, Government of India, 1983.
BBC Handbook, BBC London, 1973.
Bokhari, A.S., *Handbook for Junior Programme Staff*, for Departmental use, 1951.
Bokhari, Z.A., *Sarguzasht*, Manzoor Press, Lahore, 1995.
Constitution of Pakistan, The, 1973.
Facts about Germany, Societats-Verlag, Frankfurt am Main, Germany, 1995.
Fatmi, Prof. Hasan Askari, *Pakistan Iblaghiat*, Maktaba-i-Sahafat, Karachi, 1995.
Gazette of Pakistan Extraordinary, Islamabad, 27 March 1974.
Gilani, Saleem. A., *In Tune with National Ideals*, printed article, 1991.
Hasan, Khalid, *Zahur Azar's Day*, article in *The Nation*, 12 October 1997.
Head, Sydney W. and Sterling, Christopher H., *Broadcasting in America, A Survey of Television, Radio and New Technologies*, 4th Edition, Houghton Mifflin Co., Boston.
Jab Tarikh Ban Rahi Thi, article in *Nawa-i-Waqt*, Rawalpindi, 22 January 1988.
Keith, Michael C. and Krause, Joseph M., *The Radio Station*, Focal Press, Boston, 1986.
Legislative Assembly of India Debates: 1931-47, Delhi.
Luthra, H.R., *Indian Broadcasting*, Publications Division, Ministry of Information and Broadcasting, Government of India, 1986
Mohammad, Nauman Bin, *The Future is not for Us to See*, article in *The News*, 23 June 1995.
Nasir, Agha, Paper on 'Role of Radio in Taking the Message to the Masses', July 1992.
Nasir, Agha, Paper on 'The Evolution of Stage Play in Pakistan', July 1990
NHK, Handbook, 1970 Radio and TV Culture Research Institute, NHK, Tokyo.
NHK, *This is NHK 1975-1976*. Public Relations Bureau, NHK, Tokyo, 1975.
Pakistan-1953: Sixth Year, Pakistan Publications, Karachi.
Pakistan-1967, Pakistan Publications, Karachi.
Pakistan-1988, an Official Handbook, Directorate General of Films and Publications, Government of Pakistan, Islamabad.
Pakistan-1995, an Official Handbook, Directorate General of Films and Publications, Government of Pakistan, Islamabad.
PBC, *50th Anniversary of Radio Pakistan Lahore: 1937-1987*, Pakistan Broadcasting Corporation, Lahore, 1987.
PBC, *External Services: 26th Anniversary*, External Services, Pakistan Broadcasting Corporation, Karachi, 1976.
PBC, *Silver Jubilee of Central Productions:* 1960-85, Pakistan Broadcasting Corporation, Karachi, 1985.

PBC, YEAR-BOOK: 1973-74, 1974-75, 1975-76, 1976-77, 1977-78; Pakistan Broadcasting Corporation, Rawalpindi.

Quaid-i-Azam: Speeches and Statements 1947-48, DFP Ministry of Information and Broadcasting, Government of Pakistan, Islamabad 1989.

Radio Pakistan/PBC, Fortnightly/Monthly *Ahang* and *Pakistan Calling*.

RP, *Conversion of Radio Pakistan into a Statutory Corporation*, PC-I Form, Radio Pakistan, Rawalpindi, 1972.

RP, *Technical Achievement of Radio Pakistan: 1947-72*, Radio Pakistan, Rawalpindi, March 1972.

RP, *Ten Years of Development: 1958-68*, Radio Pakistan, Karachi, 1968.

RP, *Three Years of Radio Pakistan: August 1947 To August 1950*, Public Relations Directorate, Radio Pakistan, Karachi.

Saeed, Qazi A., Paper on *Radio Pakistan's Role in National Affairs*, November 1996.

Sham, Mahmood, article in daily *Jang*, Rawalpindi, 26 March 1989.

Shariff, Maqbul, News Story on Broadcasting Committee, *Pakistan Times*, Rawalpindi, 27 August 1967.

Siddiqi, Zamir, *Science Broadcasting, A Manual for Planning and Production of Science Programmes*, 1988.

Siraj, Syed Abdul, article on *Technological Trends in Communication and Mass Media*, 1995.

Sterling, Christopher H. and Kittross, John M., *Stay Tuned, A Concise History of American Broadcasting*, Wadsworth Publishing Co., Belmont California, 1978.

Ten Years of Pakistan: 1947-57, Pakistan Publications, Karachi.

USIS, *40 Years of USIA*, USIS, Islamabad, August 1993.

VOA Today: Office of Audience Relations, VOA, USIA, Washington, 1986.

VOA, *50 Years of Broadcasting to the World, 1942-92*, Voice of America, Washington D.C., USA.

Year-Book: 1963-64, 1964-65, 1965-66; Ministry of Information and Broadcasting, Government of Pakistan, Karachi/Rawalpindi.

Zuberi, Jamil, *Sindh's First Radio Station*, article in the *DAWN*, Karachi, 28 November 1986.

INDEX

A

A Tragedy in Focus, 235
A Treatise on Electricity and Magnetism, 1
A.Q. Laboratories, 85
Abbasi, Nilofer, 159
Abbasi, Qamar Ali, 258
Abid, Syed Abid Ali, 74
ABU, 194, 217, 224, 236, 237, 243, 256
Afghanistan, 9, 108, 110, 116, 131, 170, 171, 262
Africa, 4, 5, 94, 96, 221
Agricultural Development Bank of Pakistan, 90
Ahang, 62, 69, 71, 152, 174, 191, 205, 231, 232, 233, 234
Ahmad, Agha Bashir, 24, 180, 181, 182
Ahmad, Dr Ashfaq, 85
Ahmad, Iftikhar, 160
Ahmad, Maulana Salahuddin, 67
Ahmad, Saleem, 75
Ahmad, Shakil, 120
Ahmad, Syed Rashid, 24, 74
Ahmed, Qari Akhlaq, 58
Ahmed, Rafiquddin, 10
AIBD, 220, 221, 223, 224, 236, 237
Akashvani, 8
Al-Azhari, Qari Khushi Muhammad, 215
Ali, Chaudhry Muhammad, 80
Ali, Mahmood, 159
Ali, Mujahaid Mubarak, 160
All India Radio (AIR), 6, 7, 8, 9, 10, 11, 12, 18, 32, 87, 97, 108, 110, 117, 123, 124, 125, 180, 153, 155, 161, 209, 253, 226
All Pakistan Newspaper Society, 233
Allama Iqbal Open University, 212
AMIC, 236
Amritsari, Ishaq, 67

Ansari, I.A., 220
Ansari, K.S.H., 24
Armstrong, Edwin, 69
Arts Council, 75
ASBO, 236
Ashk, Upendra Nath, 74
Ashraf, Agha Muhammad, 21
Asia, 2, 4, 5, 94, 96, 220
Asian Games, 236
Asia-Pacific Broadcasting Union, 221, 224, 243
Asia-Pacific Institute for Broadcasting Development, 98, 220
Associated Press of Pakistan (APP), 30, 128
Atomic Energy Agricultural Research Centre, 86
Australia, 2, 5, 16, 49, 86, 94, 125, 237
Awaz Khazana, 217
Awaz, 231
Azaat-i-Pakistan, 113, 232
Azar, Zahur, 12
Azhar, Aslam, 120
Aziz, S.A., 23, 24, 182

B

Bagh-o-Bahar, 75
Bakhsh, Husain, 14
Balighuddin, Shah, 159
Baloch, Sikandar, 160
Bangladesh, 13, 112, 125, 162, 221, 237
Bano, Iqbal, 254
Batalvi, Ejaz, 67
Bedi, Rajinder Singh, 74
Beg, Mirza Qaleej, 259
Behzad, Anwar, 120
Betar Jagat, 231
Bhagwat Geeta, 54
Bhatti, Kamran, 160

Bhatti, Rashid Ahmed, 14, 23, 180
Bhavani, S.R., 14
Bhitai, Shah Abdul Latif, 80
Bhutan, 162, 221
Bible, 54, 59
Bismil, Aftab Ahmad, 12
Bokhari, A.S., 8, 9, 23, 25, 74
Bokhari, Z.A., 16, 21, 24, 58, 67, 73, 74, 170, 180, 181, 216, 257
Bolivia, 96
Branly, Edouard, 1
Britain, 4
British Broadcasting Corporation (BBC), 3, 4, 5, 8, 9, 32, 50, 54, 117, 123, 125, 153, 155, 161, 219, 223, 229, 231, 262
British Council, 223
British India, 6, 11, 108
Buddha, 179
Bulhe Shah, 80
Bureau of Agricultural Information, 92
Butt, Shamsuddin, 75, 209, 255

C

Cairo, 155
Canada, 2, 3, 49, 92, 94, 119, 141, 237
Carrapiett, Edward, 120
CBA, 224, 236
CCIR, 194, 236
CDA, 239
CDWP, 193
Central Asia, 112, 145
Central Asian Republics, 221
Chaghi, 174
Chander, Krishan, 74
Cheema, Muhammad Anwar, 160
China, 112, 175, 221, 237
Choudhry, Prof. Munir, 44
Chowdhary, Justice Abu Sayeed, 81
Civil Defence Academy, 237
CNN, 125
Cuba, 4

D

Daudpota, Allama Umar ben Muhammad, 259
de Forest, Lee, 2

Dehlavi, Dagh, 67
Dehlavi, Shahid Ahmad, 61
Denmark, 233
DESTO, 85
Deutsche Welle, 5, 220, 221, 223
Dominican Republic, 96
Doordarshan, 125
Durak, Jam, 80, 264
Dxer, 114, 230

E

East Africa, 119
East Asia, 5
East Germany, 4
East Pakistan, 13, 16, 22, 27, 61, 80, 90, 91, 95, 104, 111, 118, 155, 182, 214, 226, 235
Eastern Europe, 4
EBU, 194, 236
ECNEC, 193
ECO, 236
Edhi Welfare Centre, 40
Educational Broadcasts, 93
Effandi, Hasanali, 259
Egypt, 67, 110
Ejaz, Shamim, 120
El Khatib, H.E. El Sayyid Abdul Hamid, 58
Elan, 232
EMI, 70, 166, 214
Empire Service, 4
England, 3, 9, 86, 131
English Channel, 2
Eritrea, 110
Europe, 2, 4, 5, 112, 131, 190

F

Faiz, Faiz Ahmad, 67, 156
Far East, 4, 131
Farid, Baba Ghulam, 80
Farid, G.K., 13, 24, 180
Federal Public Service Commission, 197
Fessenden, Reginald, 2
Fielden, Lionel, 8, 32
First World War, 2
Fleming, Sir Alexander, 85

INDEX

France, 2, 4
Friedrich Ebert Stiftung, 237

G

Gallup Pakistan, 228
Gandhi, Mohandas Karamchand, 21
Germany, 2, 4, 5, 15, 86, 220, 237
Ghalib, Mirza, 179
GHQ, 103, 154
Ghulamali, Anita, 120
Gilani, Syed Saleem A., 34, 37
Girl Guides, 100, 102
Goethe Institute, 223
Gorakhpuri, Majnun, 156
Greece, 86
Gulf States, 233
Gulf, 110, 113, 144, 190
Guru Nanak, 179

H

Hafeez, S.A., 181
Haider, K., 58
Hamadani, Mustafa Ali, 12
Hameed, Akhtar, 44
Hameed, Khalid, 160
Haq, Raja Zafarul, 255
Haq, Saeedul, 24
Haque, Capt. Ahsanul, 13, 181
Haque, Sheikh Ihsanul, 24, 181
Hasan, Mumtaz, 43
Hashimuddin, Begum Mariam, 44
Hashmey, N.H., 24
Hashr, Agha, 74
Hasrat, Chiragh Hasan, 10
Herrold, Charles D., 2
Hertz, Heinrich, 1
Hla Maung, 160
Holland, 15
Honduras, 96
Hong Kong, 233
Hoshiarpuri, S.A. Hafeez, 24, 60
Hoso Banka, 217
Husain, Ch. Akbar, 58
Husain, Dr Akhtar, 10
Husain, Intizar, 75
Husain, Madho Lal, 80

Husain, Sain Akhtar, 160
Husain, Syed Sajjad, 81
Husain, Tahir, 13, 14
Hussain, Syed Munir, 91
Hyder, S.K., 14

I

Idnani, T.N., 14
IFRB, 194, 236
IIEC, 23
India, 6, 11, 13, 15, 17, 18, 19, 20, 26, 28, 33, 64, 68, 74, 87, 92, 108, 111, 117, 155, 156, 206, 221, 237
Indian Broadcasting Company, 6, 7, 209, 231
Indian State Broadcasting Service, 7, 8, 231
Indonesia, 245
Integrated Rural Development Programme, 223
International Advertising Association, 236
International Centre of Theoretical Physics, 81
International Congress on Amir Khusrau, 159
International Congress on Quaid-i-Azam, 157
International Folk Music Council, 236
International High Frequency Broadcasting, 22, 23
International Institute of Communication, 236
International Seerat Conference, 57
International Telecommunication Union, 148
Iqbal, Afzal, 181
Iqbal, Allama Muhammad, 66, 81, 88, 94, 99, 106, 212, 215, 235, 264, 158, 170
Iqbal, Dr Javed, 81
Iqbal, Nighat, 160
Iran, 15, 88, 131, 221, 233, 237
Iraq, 110
ISBO, 236
Islami Kahanian, 235
Islamiat, 95, 235
Islamic Summit Conference, 156, 215
ISPR, 103

Italy, 2, 4, 81, 86, 92
ITU, 149, 194, 236

J

Jakarta, 155
Jalal, A. Ghani Eirabie Hamid, 24, 181
Jallandhri, Abul Asar Hafeez, 156
Jamal, Begum Mumtaz, 14
Japan, 4, 5, 9, 49, 86, 92, 94, 108, 125, 168, 217, 229, 237
Jinnah, Quaid-i-Azam Mohammad Ali, 11, 13, 14, 20, 21, 38, 71, 81, 88, 106, 157, 158, 214, 215, 234
Jordan, 110, 125
Junejo, Muhammad Khan, 260

K

Kamin, Jaffar, 244
Kasi, Agha Muhammad, 267
Kazmi, Hasan Zaki, 160
Kenya, 15, 233
Khabarnama, 87, 105, 120, 121, 230
Khan, Amir, 159
Khan, Dr Abdul Qadeer, 85
Khan, Fateh Ali, 160
Khan, General Ayub, 28, 259
Khan, Hameed Ali, 160
Khan, Liaquat Ali, 16, 21, 38, 80, 88
Khan, Malik Aziz, 224
Khan, Nasrullah, 75
Khan, Nawabzada Sher Ali, 254
Khan, Nusrat Fateh Ali, 160
Khan, Pathane, 254
Khan, Riaz Ahmad, 15, 23, 24, 120, 180, 181
Khan, Younus, 224
Khatak, Khushhal Khan, 80
Khattak, Aslam, 7
Khurshid, S., 58
Khusat, Irfan, 159
Khusrau, Amir, 60, 106, 158, 215
King, Peter, 50
Kiribati, 224
KMC, 16
Korea, 244

L

Lahore High Court, 164
Latin America, 2, 4, 5, 96
Lebanon, 110
Lesotho, 96
Libya, 110
Linguistic Survey of India, 9
Lodge, Sir Oliver, 2
Loomis, Mahlon, 1

M

Madras Presidency Radio Club, 6
Maghmoom, Abdullah Jan, 12, 159
Majid, Abdul, 75
Malaysia, 98, 224, 233, 236, 237, 244, 245
Maldives, 162, 221
Malihabadi, Josh, 156
Manto, Saadat Hasan, 74
Marconi Company, 2, 3, 7, 261
Mast Twakkali, 80
Maududi, Maulana Abul Ala, 20
Maxwell, James Clark, 1
Mehdi Hasan, 178
Mehnaz Begum, 159, 160
Mexico, 2, 23
Middle East, 112, 113, 114, 119, 131, 144, 157, 190
Mirza, Anis, 120
Mirza, Wirasat, 120
Moin, Hasina, 75
Moinuddin, Khwaja, 74
Morocco, 110
Mountbatten, Lord Louis, 11
Mubarakmand, Dr Samar, 85
Mufti, Mumtaz, 156
Muhammad Ali, 81
Muhammad bin Qasim, 74
Multanekar, Surayya, 254
Mumtaz Hasan Broadcasting Committee, 34, 59, 73, 78, 93, 99, 101, 102, 103, 127, 229, 235
Musalman Mosiqar, 235
Musharraf, General Pervez, 122
Music Programmes, 59
Myanmar Service, 112
Myanmar, 109, 111, 113, 117, 232

INDEX

N

NAM, 109
NARC, 85
Naseem, Hameed, 67
Nasir, Agha, 75
Nasri, Ansar, 60, 75
National Institute of Public Administration, 237
National Security Council, 239
Naumani, Fakhr-i-Alam, 160
Nazimuddin, Khwaja, 22
Nehru, Pandit Jawaharlal, 10, 11
Nepal, 111, 162, 221, 224, 237
Nespak, 239
Netherlands, 2, 3, 4, 125, 237
New Zealand, 2, 5, 92
News and Current Affairs, 122
Niazi, S.S., 13
Niazi, Sajjad Sarwar, 13, 24, 219
Nippon Electric Company, 46
Nishtar, Sardar Abdur Rab, 13, 14
Nizami, Mahmood, 24, 60
Noor Jahan, 67, 178, 215, 217
North Africa, 112, 117, 145
North America, 5
North and Central America, 5
Norway, 233

O

Oceania, 2, 221
OIC, 236
Olympic, 236
Overseas Pakistanis Foundation, 114

P

Pakistan Agricultural Research, 90
Pakistan Athan, 113, 232
Pakistan Atomic Energy Commission, 85
Pakistan Broadcasting Academy, 185, 198, 200, 201, 220, 239
Pakistan Broadcasting Corporation (PBC), 30, 31, 34, 39, 50, 51, 52, 53, 54, 55, 56, 57, 58, 62, 63, 70, 72, 77, 79, 83, 84, 87, 88, 89, 91, 92, 98, 99, 100, 104, 106, 108, 109, 110, 112, 140, 145, 159, 173, 183, 184, 185, 187, 189, 191, 197, 198, 200, 201, 202, 204, 205, 206, 221, 235, 238, 239, 240, 243, 244, 245, 246, 252, 255, 256, 262, 265, 266, 267
Pakistan Broadcasting Foundation, 238
Pakistan Broadcasting Service, 12, 18, 129, 253
Pakistan Calling, 68, 113, 191, 231, 232, 233, 234
Pakistan Engineering Council, 85
Pakistan Institute of Management, 237
Pakistan Post Office Department, 204
Pakistan Science Conference, 85
Pakistan Science Foundation, 85
Pakistan Television Corporation, 49, 147
Pakistan Women Lawyers Association, 101
Palestine, 110, 156
Papua New Guinea, 96, 224
Patel, Sardar, 10, 11
PBC Act of 1973, 109, 183
PCSIR, 85
Performing Arts Society, 68
Pervez, Shama, 75
Philippines, 2, 237
PINSTECH, 85
Plays and Features, 73
Popoff, Alexander, 1
Prasad, Dr Rajendra, 10
Press, Public Relations and Publications, 200
PSI, 194
PTCL, 239
PWD, 198

Q

Qasmi, Ahmad Nadeem, 12, 75, 156, 178
Qasmi, Qari Waheed Zafar, 159
Qasmi, Qari Zahir, 58
Qazi, Allama I.I., 259
QSL cards, 114, 230
Quaid-i-Azam University, 217
Qudsia, Bano, 75, 159
Quran, 16, 21, 54, 55, 56, 58, 112, 114, 152, 215, 216, 218, 235
Quran-i-Hakeem aur Hamari Zindagi, 55, 235

INDEX

Qureshi, A. Sami, 265
Qureshi, Abdul Hayee, 160
Qureshi, Dr I.H., 58, 81
Qureshi, Masud, 269
Qutb, Asnain, 14, 24, 181

R

Radio Ankara, 117
Radio Australia, 117, 155
Radio Bangladesh, Dhaka, 111
Radio Beijing, 155, 229
Radio Canada International, 237
Radio Club of Bengal, 6
Radio Club of Bombay, 6
Radio Damascus, 117
Radio Dubai, 125
Radio Egypt, 56
Radio Free Europe, 4
Radio Japan, 117, 171, 229
Radio Kabul, 89, 124
Radio Marti, 4
Radio Moscow, 117, 155
Radio Pakistan Planning Board, 43
Radio Pakistan, 12, 14, 16, 17, 20, 21, 22, 23, 24, 27, 28, 37, 38, 39, 40, 41, 43, 44, 45, 46, 47, 48, 49, 50, 52, 54, 56, 57, 58, 59, 62, 63, 64, 66, 67, 69, 70, 74, 75, 78, 81, 82, 86, 87, 88, 93, 94, 96, 97, 98, 99, 100, 101, 103, 105, 106, 110, 112, 116, 117, 120, 123, 127, 128, 129, 136, 147, 148, 152, 153, 154, 155, 156, 157, 159, 162, 163, 166, 167, 170, 171, 172, 173, 175, 176, 177, 178, 179, 180, 181, 182, 197, 200, 203, 209, 212, 213, 214, 216, 220, 226, 230, 232, 233, 234, 235, 236, 241, 242, 243, 245, 247, 250, 251, 254, 257, 258, 266
Radio Peking, 117
Radio Shanghai Music Festival, 254
Radio Tehran, 117
Rafi Peer, 74, 75
Rafique, Muhammad, 160
Rahman, M.A., 182
Rahman, M.S., 181
Rahman, Qari Obaidur, 159
Rahmani, Ishrat, 74
Raipuri, Akhtar Husain, 156

Rajagopalachari, 10
Ramayana, 54
Rashed, N.M., 24, 181
Rasool, Qari Ghulam, 215
Raushni, 53, 55, 233, 235
Rehman Baba, 80
Rehman, Dr Attaur, 85
Rehman, M.A., 24
Rehman, Qari Abdul, 58
Reith, Lord, 32
Religious Programmes, 54
RIAS, 4
Riazul Haq 160
Rickard, J.K., 50
Riffat, Syed Ahmad, 75
Rizvi, Kamal Ahmad, 75
Roshanara Begum, 156
Ruet-i-Hilal Committee, 55
Rural and Farm Broadcasts, 88
Russia, 4

S

SAARC, 109, 142, 145, 160, 161, 218, 223, 227, 256
Sachal Sarmast, 80
Sada-i-Pakistan, 113, 232
Salam, Dr Abdus, 81, 85, 120
Salman, Ahmad, 13, 24, 180, 181, 254
Sarang, 231
Sarfraz, Mohamad, 24, 180, 181
Sarnoff, David, 2
Sarwar, Azim, 160
Saudi Arabia, 15, 57, 58, 110, 157, 174
Sautul Quran, 53, 57, 252
Sayeed, Jahanara, 120
Sayeed, M.M., 24
Science International, 80, 86
Second World War, 4, 9, 23
Secretariat Training Institute, 237
Sethi, Yunus, 12
Shabds, 54
Shad, Ata, 159
Shafi, Maulana Mufti Muhammad, 55
Shah, Wajid Ali, 60
Shahabuddin, Khwaja, 16, 104
Shahnama, 152

INDEX

Shalimar Recording and Broadcasting Company, 70, 214
Sham, Mahmood, 40
Shankar, 11
Sharqi, Sultan Husain, 60
Shindur Polo Tournament, 264
Shuja, Hakeem Ahmad, 75
Siddiqi, Dr Raziuddin, 67, 85
Siddiqi, Dr Salimuzzaman, 85
Siddiqi, Inam, 160
Siddiqi, Siddique Ahmad, 21
Siddiqui, Tanwir, 160
Siddiqui, Zamir, 224
Singapore, 5, 237, 244, 245
Singh, Sardar Baldev, 11
Soomro, Munir Ahmad, 260
South Asia, 5, 29, 69, 98, 174, 190
South Asian Federation, 160
South Asian Service, 111
South East Asia, 112, 113, 118, 119, 131
Soviet Union, 2, 4
Spain, 2
Star Plus, 125, 162, 221, 237
Stubblefield, Nathan B., 2
Sudan, 110
Suez Canal, 110
Sulemani, Feroz, 14
Sumar, A.K., 44
Supreme Court, 164
Swaziland, 96
Sweden, 86
Switzerland, 125
Syed, B.H., 14
Syed, M.H., 79
Syed, Sajida, 159
Syria, 110

T

Tabassum, Sufi Ghulam Mustafa, 67, 156
Taj, Syed Imtiaz Ali, 44, 73, 74
Tarannum Naz Laila, 178
Tarannum, 68
Taubat-un-Nosuh, 75
Ten Years of Development, 235
Thailand, 237, 245
Thanvi, Maulana Ehtishamul Haq, 54, 58, 215

The Indian Listener, 231
The Indian Radio Times, 231
The Times of India, 6
Third Conference of Station Directors of Radio Pakistan, 27
Third World, 35, 109, 190
Thomson Television International, 46
Three Years of Radio Pakistan, 234
Tipu Sultan, 74, 105, 175
Tirmizi, Ahmad Sajjad, 256
Tripitaka, 59
Tunisia, 110
Turkey, 3, 88, 111, 131
Turkish Radio and Television Corporation (TRT), 3
Turkish Wireless Telephone Company, 3
Twenty Years of Radio Pakistan, 234

U

Umar, Maulana Muhammad, 217
Umme Kulsum, 67, 215
UNDP, 76
UNESCO, 94, 148, 220
UNFPA, 76
UNICEF, 224, 256
United Consultants Limited, 50
United Kingdom, 15, 114, 233, 237
United States of America, 1, 15, 16, 86, 141, 237
USIS, 68, 86, 224, 229
Usmani, Dr I.H., 85
Usmani, Maulana Shabbir Ahmad, 16
USSR, 171

V

Vanshi, Dr Yadu, 10
Vatsayana, S.H., 10
Voice of America (VOA), 4, 5, 54, 117, 123, 125, 155, 229
Voice of Indonesia, 117

W

Waheed, Dr A., 44
Wali Dakni, 60
WAPDA 195, 197

Waqt Ki Awaz, 20
Warner, Eric, 120
Wasti, Rizwan, 120
West Africa, 155
West Pakistan, 13, 18, 27, 61, 80, 90, 91, 111, 117, 118, 124, 129, 133, 134, 136, 137, 182, 226, 258
Western Europe, 119
Wireless Telegraphy Act of 1933, 204, 205
Wireless Telegraphy Act, 14
Women National Guards, 100
World Radio Conference, 22

Y

Yemen, 110
YMCA, 7, 253

Z

Zaid, Shaista, 120, 160
Zaidi, B.H., 14
Zee 5, 125
Zee TV, 125
Ziaul Haq, President, 262